The Terry Lectures

Master Control Genes in Development and Evolution

Other volumes in the Terry Lecture Series available from Yale University Press

Becoming: Basic Considerations for a Psychology of Personality
Gordon W. Allport

Belief in God in an Age of Science
John Polkinghorne

A Common Faith
John Dewey

The Courage to Be
Paul Tillich

Education at the Crossroads
Jacques Maritain

Freud and Philosophy: An Essay on Interpretation
Paul Ricoeur
translated by Edward Quinn

The Meaning of Evolution: A Study of the History of Life and of Its Significance for Man
George Gaylord Simpson

Psychoanalysis and Religion
Eric Fromm

Psychology and Religion
Carl Gustav Jung

Master Control Genes in Development and Evolution: The Homeobox Story

Walter J. Gehring

YALE UNIVERSITY PRESS
NEW HAVEN AND LONDON

Designed by Gregg Chase and set in Minion type by Rainsford Type.
Printed in the United States of America by Vail-Ballou Press, Binghamton, New York.

Library of Congress Cataloging-in-Publication Data
Gehring, Walter J., 1939–
Master control genes in development and evolution : the homeobox story / Walter J. Gehring.
p. cm. — (The Terry lectures)
Includes bibliographical references and index.
ISBN 0-300-07409-3 (alk. paper)
1. Homeobox genes. 2. Developmental cytology. 3. Evolutionary genetics. I. Title. II. Series.
QH447.8.H65G44 1998
572.8'6—dc21 98-19922
CIP

A catalogue record for this book is available from the British Library.

The paper in this book meets the guidelines for permanence and durability of the Committee on Production Guidelines for Book Longevity of the Council on Library Resources.

10 9 8 7 6 5 4 3 2 1

The Dwight Harrington Terry Foundation Lectures on Religion in the Light of Science and Philosophy

The deed of gift declares that "the object of this foundation is not the promotion of scientific investigation and discovery, but rather the assimilation and interpretation of that which has been or shall be hereafter discovered, and its application to human welfare, especially by the building of the truths of science and philosophy into the structure of a broadened and purified religion. The founder believes that such a religion will greatly stimulate intelligent effort for the improvement of human conditions and the advancement of the race in strength and excellence of character. To this end it is desired that a series of lectures be given by men eminent in their respective departments, on ethics, the history of civilization and religion, biblical research, all sciences and branches of knowledge which have an important bearing on the subject, all the great laws of nature, especially of evolution . . . also such interpretations of literature and sociology as are in accord with the spirit of this foundation, to the end that the Christian spirit may be nurtured in the fullest light of the world's knowledge and that mankind may be helped to attain its highest possible welfare and happiness upon this earth." The present work constitutes the forty-fourth volume published on this foundation.

Contents

Foreword

Frank Ruddle

The Terry Lectures were founded in 1924 with a gift from the Dwight Harrington Terry Foundation. The lectures are intended to "nurture . . . the Christian spirit . . . in the fullest light of the world's knowledge." Many titles in the series are religious in character, but a substantial number deal with science in an effort to reconcile the tensions that exist between the two. Darwin and his revolutionary views were clearly in the minds of the founders. The deed of gift states that although religious concerns should be the focus, "all the great laws of nature, especially of evolution," be included.

This book recounts one of the major discoveries affecting our view of ourselves as human beings. The selection committee members for the 1993 lectures reproduced here were unanimously enthusiastic for the selection of Walter Gehring as lecturer and author of a related book because he has made and continues to contribute to one of the major scientific breakthroughs of this century. Walter Gehring provided the key to an understanding of how multicellular organisms, ourselves included, undergo the developmental journey from fertilized egg to fully formed individual. These findings, moreover, promise a deeper explanantion of the mechanisms regulating evolution. Certainly, Gehring's story would be consistent with the broader aims of the Terry Lectures: our understanding of self and our place in the universe.

Walter Gehring has a good ear for scientific experiment. I came to know him when we were both junior professors at Yale teaching developmental biology. A good science teacher knows how to identify the nub of an experiment and reveal it to the student in all its glory or shortcomings. Gehring reveals this talent in his book. I was again to enjoy his capacity for experimental design during a half-year stay in his laboratory in Basel in 1983, when the homeobox—what this book is all about—was discovered. During that short period, the homeobox, a 180-nucleotide stretch of DNA, led us to an understanding of a pattern of master control genes in our bodies, limbs, hands, heart, and brain.

The discovery of the homeobox expanded our self-awareness in two principal ways. First, master control genes provided an awareness of the overall genetic plan that controls the development of organisms from egg to adult. Second, a survey of all the major multicellular animal groups has shown that all contain a similar and related set of master control genes, revealing the astounding fact that during a billion and more years of evolution, a single maturing system of developmental control has produced the variety of biological life that we know today—flies, snails, sea stars, fishes, and ourselves!

This book recounts a great scientific "Genesis" story and much more. In simple and unerring economy of language Gehring teaches the basics of modern molecular biology in terms that the interested general reader can understand. He treats us to a historical account of the growth of the developmental sciences. Gehring is especially good at drawing insightful vignettes of influential investigators and mentors past and present. Walter unabashedly counts himself among them, and it is refreshing to have the self-evaluation of one of our outstanding scientists in such candid and unself-conscious terms. To read this book is to enter into an understanding of one of the great intellectual adventures of our times. It resounds to the words of the Terry bequest, "that the Christian spirit may be helped (through knowledge) to attain its highest possible welfare and happiness upon this earth."

Preface

This book describes the journey of a scientist into the vast area of developmental biology over more than thirty years. It offers a guided tour from odd mutations of the little fruit fly *Drosophila melanogaster,* transforming the antennae on the head of the fly into legs, to the isolation of the respective gene that causes this homeotic transformation, and to the discovery of the homeobox. This small DNA segment characteristic for such homeotic genes is found not only in insects but universally in all higher organisms, including humans, and provides a key to understanding development. Homeotic genes have been followed from the antennal legs down to the molecular and even atomic level by determining the three-dimensional structure of the homeodomain, which the homeobox encodes. Finally, the discovery of a master control gene for eye development is discussed with its wide implications for the evolution of this organ of extreme perfection. This book shows how much of the developmental program is written into our genes.

Of course, the pathway of this scientific journey was not as straightforward as outlined in retrospect but rather contorted with numerous sidelines leading in the wrong direction or terminating in dead ends that are largely omitted here. The journey of a scientist is perhaps best described by something that happened to me at a meeting in Point Reyes, right on the San Andreas Fault in California. The meeting, entitled "From Phage to Drosophila," was dedicated to my friend David Hogness, a pioneer in this field. After a long day of scientific discussion, I returned to my cabin and noticed a big squashed fly on the white painted wall of my bedroom. My first thought, betraying my Swiss origins, was to remove this ugly black thing from the wall. But my thoughts soon drifted away from the fly to more profound scientific problems, and I eventually forgot about it and went to sleep. The next morning, much to my surprise, the fly was now about thirty centimeters higher up on the wall. This was a clear violation of the physical law of gravity, which ought to apply at any place on earth, even along the San

Andreas Fault. Because I was sure that I hadn't had too much to drink the night before, I was convinced that the squashed fly was indeed higher up on the wall. And so I took a closer look at the fly and discovered that a tiny ant was pulling this huge prey up the wall toward its nest. This must have been the biggest catch the ant had ever made, at least ten times bigger than the little ant itself. In eight hours the ant had moved its prey by thirty centimeters only. This put me in mind of the developmental scientist's quest. Though the fruit fly is tiny, the problem it poses is gigantic. Like the small ant, scientists move the fly step by step; some lift it perhaps by thirty centimeters per night, others by five centimeters only, and some drop it altogether. At the end of this book readers may judge for themselves how far we have moved the fly during the past thirty years.

The problem for the ant was to bring the fly back to the nest. In Ernest Hemingway's "The Old Man and the Sea," an elderly fisherman catches the biggest fish of his life after a tremendous struggle and must bring it back ashore. When he turns to sail home, however, sharks attack the giant fish, which is tied to the boat, and the old man returns with just a skeleton. To find out whether a similar disaster had befallen my ant, I returned to my room during a coffee break. Happily, there were no sharks in my room, but the cleaning woman had made up my bed and had obviously removed the fly from the wall. And so I shall never know what happened to the poor ant, but as a scientist I can certainly sympathize with it.

This book summarizes the Terry Lectures I gave in the fall of 1993 at Yale University, my second alma mater. I am grateful to the members of the board of the Dwight Terry Foundation for their invitation to present the lectures and to all my friends and colleagues at Yale for their warm reception and long years of friendship.

Master Control Genes in Development and Evolution: The Homeobox Story is a very personal account that was written in part during the summer of 1994 when I was a Darwin Fellow at the University of Edinburgh. I am grateful to David Finnegan and to Noreen and Ken Murray for their kind hospitality in Edinburgh. The bulk of the manuscript stems from my stay on the Greek island of Skyros, where Popy and Spyros Artavanis-Tsakonas loaned me their beautiful house on the beach as a hideaway. Without their generosity this book may never have been finished. My thanks also go to my wife, Elisabeth, for her patience, her feedback, and her constructive criticism. I am indebted to the editors at Yale University Press, Jean Thomson Black and

Laura Jones Dooley, for removing my insults to the English language from the manuscript and polishing my style. My gratitude also goes to my secretary, Erika Marquardt-Wenger, for patiently typing the manuscript, to Margrit Jäggi, Verena Grieder, and Liselotte Müller for preparing the figures, and last but not least to all the members of my group for their great contributions and excellent teamwork.

Individual chapters were kindly reviewed by my friends and colleagues Max Birnstiel, Eddy De Robertis, Denis Duboule, Corey Goodman, Ueli Grossniklaus, Peter Gruss, Ernst Hafen, Herbert Jäckle, Ed Lewis, Nori Satoh, Alexander Schier, Stephan Schneuwly, Gerold Schubiger, Paul Sternberg, Eric Wieschaus, Debra Wolgemuth, and Kurt Wüthrich. Their criticism and suggestions were most helpful. I hope that those of my colleagues whose contribution is not adequately covered in this book will forgive me; it represents my personal view and is therefore certainly biased. But I hope to convey to both scientists and general readers some of the excitement generated by the elucidation of the genetic program that controls development and evolution.

Chronology

1859	Darwin	*On the Origin of Species*
1866	Mendel	*Versuche über Pflanzenhybriden*
1871	Miescher	Isolation of nuclein
1894	Bateson	Materials for the study of variations (homeosis)
1902–3	Boveri, Sutton	Chromosome theory of inheritance
1910	Morgan	*white* mutation and sex linkage in *Drosophila*
1915	Morgan, Sturtevant, Muller, and Bridges	Mechanism of Mendelian heredity
1915	Bridges	*bithorax* mutation
1924	Spemann and Mangold	Organizer experiment, embryonic induction
1933	Heitz and Bauer, Painter	Giant polytene chromosomes
1934	Morgan	Theory of differential gene activity
1944	Avery, MacLeod, and McCarty	hereditary material is DNA
1953	Watson and Crick	Double helical structure of DNA
1958	Brenner, Jacob, and Meselson	Messenger RNA
1960	Jacob and Monod	Regulatory genes, operon model
1966	Nierenberg	Genetic code
1966	Birnstiel	Isolation of ribosomal RNA genes from *Xenopus*

1973	Berg, Cohen, and Boyer	Gene cloning
1978	Lewis	Bithorax complex
1979	Khorana	Total synthesis of a gene
1980	Nüsslein-Volhard and Wieschaus	Segmentation and maternal-effect mutants
1982	Spradling and Rubin	P-element transformation in *Drosophila*
1983	Hogness	Cloning of *bithorax* gene
1984	Gehring, Scott	Discovery of the homeobox
1989	Capecchi	Gene transfer by homologous recombination in the mouse
1989	Wüthrich	Structure of the homeodomain

Master Control Genes in Development and Evolution

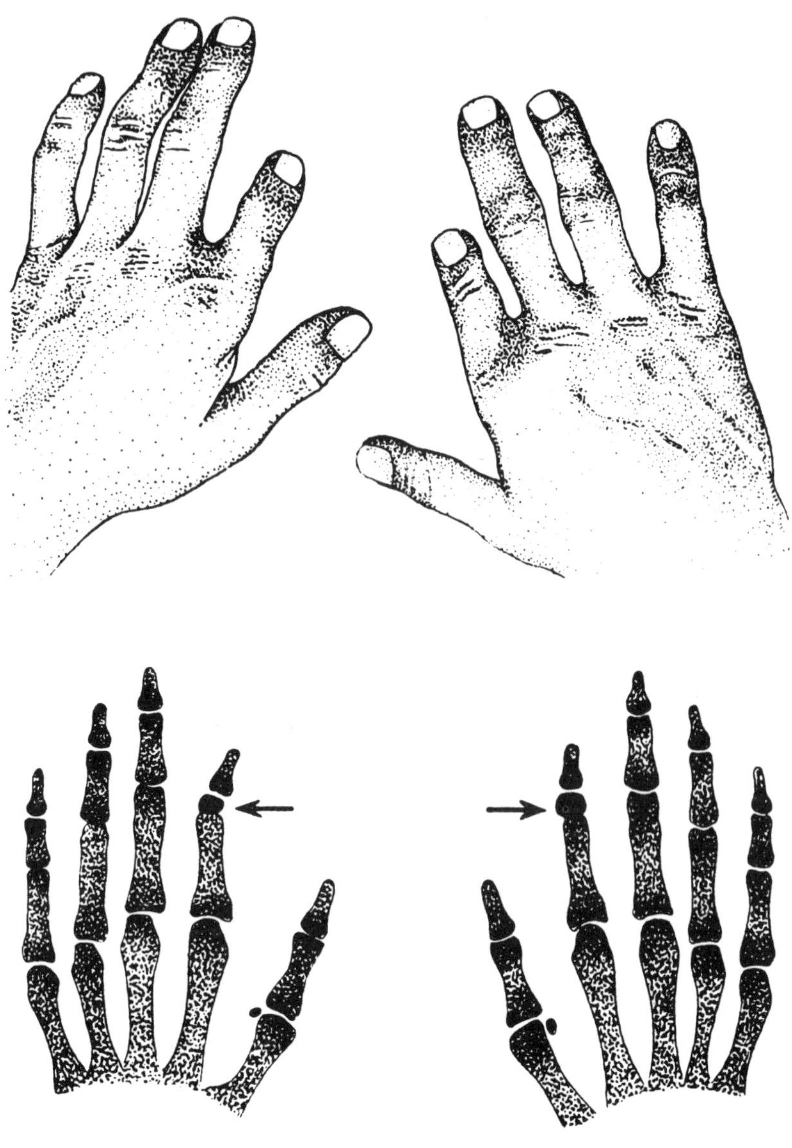

Figure 1.1
Short fingeredness ("brachydactylism"), a dominant mutation in humans that leads to a reduction of the second bone of the second digit only. From E. Hadorn, Letalfaktoren *(Stuttgart: George Thieme Verlag, 1995).*

~

1
The Ancient Language of Genes

IT WAS a hot summer day, and I was on a flight from New York to Seattle to give a talk at the University of Washington. On board I was seated next to a young woman who, judging from her English accent, might have been of Scandinavian origin. As we fastened our seatbelts, I noticed that she had unusually short second fingers on both hands. I remembered from lectures in genetics, which I had attended as a young student, that this "short fingeredness" is a heritable trait that was first described as "brachydactylism" in a Norwegian family and is caused by a dominant mutation in a single gene. A defect in this one gene reduces the second bone of the second finger (fig. 1.1), which means that the normal gene controls the size and shape of just this one bone. My mentor, Ernst Hadorn, used "short fingeredness" as an example of how our body plan, in all its details, is laid down in our genes. Understanding how this works is the subject that has fascinated me throughout my life as a biologist.

What I did not remember from Hadorn's lectures was whether this particular mutation affects only the fingers or whether the corresponding toes are also reduced in size. So I decided to have a look at my co-passenger's toes. This was not easy, however, as there is never enough leg space in tourist class. Fortunately, the young woman was wearing sandals, and I casually dropped my newspaper so that I could inspect her toes while trying to collect the paper from the floor. Indeed, I can assure you that she also had very short second toes. This observation reveals another important principle: the same gene controls homologous structures. Fingers and toes are homologous, or equivalent, structures; the second bone of the second finger is constructed according to basically the same genetic instructions used to form the second bone of the second toe. Homologous structures are specified by variations of the same genetic program. This makes development and evolution similar to a great musical composition like a fugue of J. S. Bach, which also involves variations on a basic theme. Brachydactylism is only one example showing how precisely our body plan is specified by our genes. The genome, the sum of our genetic material, provides the framework within which we develop and sets the limits for influences from the environment.

Where and how is this genetic program written down? The man who had the first clear ideas of this was Friedrich Miescher, who was a professor at the University of Basel in Switzerland during the second half of the nineteenth century (fig. 1.2). The Miescher family originally came from Bern, and his father took a professorship in physiology at the University of Basel, where Friedrich Miescher later studied medicine. Miescher was close to his uncle, Wilhelm His, a famous embryologist. Their friendship is reflected in their extensive correspondence, which spanned more than three decades.

Wilhelm His apparently suggested that his young nephew should investigate the biochemistry of the cell nucleus. Miescher accordingly went to Tübingen to work with Ernst Hoppe-Seyler to carry out what we would today call postdoctoral work. Miescher decided to use lymphocytes for his biochemical studies because these cells have a large nucleus and little surrounding cytoplasm and they are isolated rather than embedded in tissue. He recovered the cells from the bandages of patients at a nearby hospital (before the discovery of antibiotics, bacterial infections were extremely common, and the lymphocytes—pus cells—were easy to come by). The technical means available then for biochemical research were incredibly primitive. Miescher had only very crude tools to purify cell nuclei and extract their contents. To digest the cell cytoplasm, he used pork stomachs as a source for crude preparations of the enzyme pepsin, which can digest proteins and peptides. He also found that he could precipitate the major nuclear substance with alcohol and redissolve it in dilute salt solutions, but these were essen-

Figure 1.2
Friedrich Miescher, discoverer of nucleic acids. Courtesy of the Portrait Collection of the University Library, Basel.

tially all the tools he had at his disposal. Nevertheless, Miescher succeeded in isolating a substance he called nuclein, what we now call nucleic acid. This nuclein consisted largely of DNA (deoxyribonucleic acid) with some contaminating RNA (ribonucleic acid) and protein. Hoppe-Seyler was a leading chemist with a well-equipped laboratory, and there Miescher was able to determine nuclein's phosphorus content. Although nuclein behaved like a protein, consisting of numerous carbon atoms along with hydrogen and nitrogen, it differed from the known proteins in its high phosphorus content. The isolation and characterization of nuclein undoubtedly represent the discovery of DNA and are a milestone in the history of science. Miescher submitted his results to Hoppe-Seyler's *Journal medicinisch-chemische Untersuchungen* in 1869, but publication was delayed until 1871. As an editor of this journal, Hoppe-Seyler repeated all of Miescher's experiments to ensure that they were correct before he would accept the paper for publication. Times have changed!

Strangely, most historians of science are not aware of the far-reaching implications of Miescher's discovery, which represents the birth of molecular biology. Edgar Bonjour, a renowned Swiss historian who wrote a six-hundred-page history of the University of Basel, mentions the discovery of nucleic acids by Miescher in one sentence: "His first independent piece of work on the chemical composition of pus cells received much attention." Even the eminent biologist Ernst Mayr implies in *The Growth of Biological*

Thought that Miescher did not realize the importance of his discovery and therefore did not follow up on his DNA work. But this criticism is completely unjustified, as Miescher's correspondence with Wilhelm His shows. After becoming a professor of physiology at the University of Basel, Miescher continued his work on nuclein. In Basel he isolated nuclein from salmon sperm, another excellent choice, because sperm is a rich source of DNA and salmon were then abundant in the Rhine River. Unfortunately, biochemical methods of the time were totally inadequate for solving the structure of DNA, and Miescher later had to abandon the project. He could not distinguish clearly between proteins and phosphorus-rich nuclein, but he realized that nuclein had properties of a large molecule. In several letters to his uncle, Miescher formulated his prophetic and revolutionary ideas about nuclein as the genetic material.

Miescher was a strong proponent of the idea that the hereditary material is a chemical substance, which contradicted prevailing beliefs. At that time, most biologists were morphologists who believed heredity to be based on vitalistic or mystical principles. The idea that the hereditary material could be isolated in a test tube was absolutely inconceivable. But Miescher's proposal went far beyond the idea that the hereditary material is a chemical substance. He was also convinced that the hereditary substance was a single type of molecule rather than a multitude of different molecules, one for each gene. Information was stored, he proposed, in the groups of atoms that constituted the hereditary molecule, or, more precisely, in the stereochemistry of these groups of atoms. These revolutionary ideas, which were largely confirmed some sixty years later, are clearly expressed in Miescher's correspondence with his uncle. On December 17, 1892, he wrote, "For me the key to sexuality lies in the stereochemistry. The 'gemmules' of Darwin's pangenesis are nothing else than the numerous asymmetric carbon atoms in the organized substances. . . . In the enormous protein molecules (egg white bodies) or in the even more complex molecules of hemoglobin, etc., the numereous asymmetric carbon atoms allow a colossal amount of stereoisomers, so that all the richness and all variations of hereditary transmissions can find their expression equally well, as the words and terms of all languages in the 24–30 letters of the alphabet" (for the complete letter, see Appendix 1). In this prophetic correspondence, Miescher clearly preconceives the idea of a genetic code. For the first time he compares the information contained in the hereditary material to human languages, which allow us to store information in a sequence of letters of the alphabet as written text. We know today that the genetic information is contained in the sequence of the bases in the DNA rather than in the sequence of asymmetric carbon atoms, but the principle remains the same. The language of genes is quite analogous to human language.

Asymmetric carbon atoms had just been discovered by Louis Pasteur, who showed that a molecule containing such atoms, such as tartaric acid, can exist in two spatially different forms, or stereoisomers, even though they are composed of the same combination of atoms. The two stereoisomers form different crystals that are mirror images. If a single molecule contains several asymmetric carbon atoms, the number of possible stereoisomers it can acquire increases exponentially according to the formula 2^n, where n is the number of asymmetric carbon atoms. In his letter of October 13, 1893, Miescher expands on this idea and calculates the number of possible stereoisomers: "The continuity lies not only in the form, it also lies deeper than the chemical molecule. It lies in the groups of atoms that constitute the molecule. In this sense I am a proponent of the chemical theory of inheritance." Later he continues, "If, as it is easily possible, a protein molecule contains 40 asymmetric carbon atoms, this results in 2^{40} . . . isomers. . . . In order to account for the immense variability postulated by the theory of inheritance, my theory is better suited than any other." This letter (see Appendix 2) illustrates Miescher's quantitative approach. The estimated figure of 2^{40}, which equals approximately 10^{12}, or one trillion, indicates that Miescher was on the right track. His idea that genetic information is contained in the stereochemistry of a single molecule proved to be basically correct. One has only to substitute asymmetric carbon atoms with base pairs to arrive at our present understanding of how genetic information is stored.

During the nineteenth century three major theories laid the foundation of modern biology: Darwin's theory of evolution, Mendel's concept of genetics, and Miescher's hypothesis on the chemical nature of the genetic material. The integration of these three theories led to the revolutionary development of biology in the twentieth century. Charles Darwin's ideas on evolution were the most influential politically, since their implications were immediately obvious to the layperson. Darwin not only put forward an evolutionary theory but provided an explanation for the driving force of evolution in natural selection. Yet because Darwin did not distinguish clearly between hereditary and acquired characters, his evolutionary theory had a weak point: the pangenesis theory of inheritance. In Darwin's view, each part of the body produced something called gemmules, which were somehow collected in the semen (as we would say nowadays, the germ cells). These gemmules developed into the characters of the offspring and therefore were supposed to be the material basis of heredity. It was Gregor Mendel, a contemporary of Darwin, who elucidated the rules of inheritance and postulated the existence of genes, but Mendel's work was not recognized until the beginning of the twentieth century. Only after Darwin's death, then, did the neo-Darwinians integrate Mendelian genetics into Darwin's theory. Miescher, too, was well aware of the weaknesses of the pangenesis theory, and in the letter cited above he provides a concrete explanation for Darwin's gemmules by proposing that they correspond to asymmetric carbon atoms in the hereditary substance—in other words, to stereochemical differences along the genetic molecule, a revolutionary thought that was confirmed much later.

Independently of Miescher's biochemical work, Mendel discovered the rules governing inheritance, the transmission of defined characters from generation to generation. Mendel was a Catholic priest who had studied in Vienna and did his genetic experiments on plants, mainly peas, in the garden of the monastery of Brünn. Before Mendel, ideas about inheritance were rather diffuse, mainly because of the lack of understanding of fertilization. In contrast to most of his contemporaries, Mendel was convinced that both the father and the mother contribute equally to the progeny, and his experiments on peas certainly confirmed this view. His great accomplishment was that he took a quantitative reductionist approach: he considered only single, well-defined heritable traits such as the color of the flowers or the form of the seeds, and he followed the inheritance of these characters quantitatively by counting the number of progeny with a given character. He first generated pure lines that consistently showed the same character over several generations and then crossed them to each other and determined the number of progeny with a given trait or a combination of traits in subsequent generations. From these experiments Mendel was able to derive the rules governing inheritance. He also deduced the existence of genes, which he called cell elements, but he did not have the vaguest notion about their nature (incidentally, the term *genes* was coined much later in 1909 by H. Wilhelm Johannsen). Mendel's findings implied that the observed frequencies of progeny were due to random combinations of genes generated during the formation of germ cells (sperms and eggs) and during fertilization.

Fertilization is a big lottery, and the fate of the progeny is decided on the basis of which of the millions of sperm fertilizes a given egg in a largely random process. That most germ cells differ from one another with respect to one or more genes indicates that every individual, with the exception of identical twins, represents a unique combination of genes. For the individualists among us (and I am one of them) this is a comforting thought. Each one of us is a unique combination. Once fertilization has occurred, we are genetically programmed, and even though there are strong environmental influences, they are limited by the genetic framework. Mendel's experiments indicated that peas have two copies of each gene, one from the mother and one from the father, and that the genes re-assort during germ cell formation, with germ cells containing only one copy of each gene; at fertilization, the genes of egg and sperm are combined to give the double (diploid) number again.

After establishing the rules of inheritance for peas, Mendel studied a variety of other plants and found that his rules also applied to the flower stock (*Matthiola*) and maize and to some extent to animals. But there were exceptions! Following a suggestion by Karl von Naegeli, a renowned botanist in Munich, Mendel also examined hawkweed (*Hieracium*), which to his great dismay gave completely different results. This may have been one of the reasons why Mendel's rules were not immediately accepted. In retrospect, it is not surprising that *Hieracium* was an exception, since it develops without fertilization and with only maternal genes.

Once Mendel's work was rediscovered in the early twentieth century, it soon became apparent that his rules applied to both plants and animals, but elucidating the nature of genes proved more difficult. Cytologists had described threadlike structures called chromosomes that appear during cell division in the cell nucleus and can be stained with certain dyes. Each species they examined of animal and plant had a characteristic number of chromosomes, which appeared as duplicated structures before nuclear division and were distributed among the daughter cells so that each daughter nucleus received exactly one copy of each chromosome. However, during interphase, in between divisions, the chromosomes apparently disappeared and could no longer be detected. Theodor Boveri first showed that the chromosomes continue to exist through interphase in the early embryonic stages of the worm *Ascaris*, which is ideally suited for cytological studies. By studying the development of sea urchin eggs that were fertilized by two sperms instead of one, Boveri showed that only cells with a complete set of chromosomes can develop normally, whereas cells missing individual chromosomes or carrying additional chromosomes developed abnormally. On the basis of these observations, Boveri concluded that chromosomes were the carriers of genes and proposed the chromosomal theory of inheritance. Harry E. Sutton, studying the chromosomes of a grasshopper, independently arrived at the same conclusion. He found that during germ cell formation the number of chromosomes is reduced to one half (haploid number) by two so-called meiotic divisions but that the double (diploid) number of chromosomes is restored during the process of fertilization, in which the sperm and egg cells fuse. One chromosome set comes from the mother, and one set is contributed by the father. The distribution of the chromosomes corresponds exactly to the segregation behavior of the Mendelian genes.

The rigorous proof for the chromosomal theory of inheritance came from work on the fruit fly, *Drosophila melanogaster*. Thomas Hunt Morgan introduced *Drosophila* as a model organism into genetics and established an excellent group of collaborators working in the famous "fly room" at Columbia University, where a number of fundamental genetic discoveries were made. In 1910, Morgan discovered the first mutation, called *white* (*w*), which results in white eye color rather than the brilliant red observed in wild flies. It was left to Calvin Bridges, one of Morgan's outstanding students, to prove the chromosomal theory of inheritance. The *white* mutation showed sex-linked inheritance, and by isolating flies in which the sex chromosomes were abnormally distributed during germ cell formation, he showed that the *white* gene always segregated with the X-chromosome, providing definitive evidence that the X-chromosome carries the *white* gene. Another student, Alfred Sturtevant, subsequently showed by ingenious experiments that the genes are arranged in a linear order along the chromosomes and that each gene can be assigned a defined position on the chromosome.

Yet the biochemical nature of the genes remained elusive. An entry point for identifying Miescher's hereditary substance (nuclein) was provided by Fred Griffith's transformation experiments using bacteria. Griffith discovered that killed bacteria, when brought into contact with live bacteria of a different strain, could transmit a hereditary character from the dead donor to the live recipient, a phenomenon called genetic transformation. In a series of painstaking experiments, Oswald Avery and his collaborators demonstrated that DNA rather than protein was the transforming principle, and they concluded that DNA was the hereditary substance. This conclusion was supported by independent evidence from Alfred Hershey and Martha Chase, who studied bacteriophages, viruses that infect bacteria. They showed that the virus attaches to the outside of the bacterium and injects its DNA into the bacterial host cell, whereas the protein envelope of the virus remains outside. The injected DNA serves as a blueprint or template in producing large numbers of phage progeny. This experiment also led to the conclusion that the DNA, rather than the protein, contains the hereditary information passed on to the progeny. Even though both experiments did not rule out the possibility that minor contaminating molecules might be important, most molecular biologists became convinced that DNA was the hereditary substance.

This opened the competition for determining the structure of DNA, which is vividly described in James Watson's *The Double Helix.* Watson and his colleague Francis Crick, in their famous paper published in *Nature* in 1953, proposed that DNA consists of two complementary chains of bases attached to a backbone consisting of alternating sugar residues and phosphate groups (fig. 1.3). The two complementary chains are held together by hydrogen bonds between two complementary bases adenine (A) and thymine (T), forming an A-T base pair, and guanine (G) and cytosine (C), which form a G-C base pair. However, the two chains are not straight and parallel to each other but rather wound into a helix, like a spiral staircase in

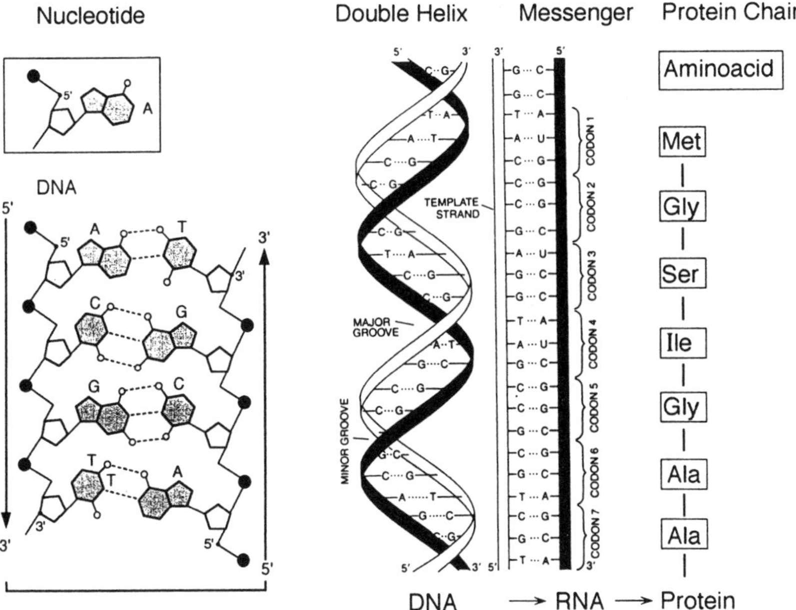

Figure 1.3
Composition and structure of DNA, RNA, and protein

which the base pairs form the steps and the backbone represents the handrail (see fig. 1.3). Further, the two chains have opposite polarity. This model suggested immediately how a DNA molecule might duplicate itself and how genetic information could be stored. As Watson and Crick wrote, "One chain is, as it were, the complement of the other, and it is this feature which suggests how the DNA molecule might duplicate itself." One chain serves as a template for the synthesis of the complementary chain. If one chain reads, for example, >*ATTGCCA*, the complementary strand has to read *TAACGGT*< in the opposite orientation, so that A always pairs with T and G with C. "It follows," they stated, "that in a long (DNA) molecule many different permutations are possible, and it therefore seems likely that the precise sequence of the bases is the code which carries the genetical information." This is exactly along the line of Friedrich Miescher, except that there are four letters in this alphabet (A, T, G, and C) rather than two as for the asymmetric carbon atoms, but the principle is the same, and of course, the high phosphorus content found by Miescher is also accommodated by this model. Both predictions of Watson and Crick were soon confirmed experimentally.

The first problem was to find out how the complementary DNA chains were generated and whether DNA replication is catalyzed by enzymes. Arthur Kornberg's biochemical experiments soon showed that enzymes, DNA polymerases, were indeed essential for replicating DNA, but it also required genetic evidence to determine which of the enzymes accomplishes this task in vivo. This complex enzyme is capable of adding the nucleotides—the building blocks, which consist of one base, a sugar moiety, and a phosphate group—one after the other onto the growing chain (see fig. 1.3) in the correct sequence as dictated by the complementary strand, which fulfills the prediction by Watson and Crick.

How is the information transferred from the DNA to protein molecules? Both DNA and proteins are linear chains consisting of building blocks, but DNA contains only four nucleotides, A, T, G, and C, whereas proteins are composed of twenty kinds of amino acids. Experiments carried out by François Jacob at the Louis Pasteur Institute in Paris first hinted at the existence of a short-lived intermediate in the information transfer. This intermediate was shown to be RNA and was designated as messenger RNA (mRNA) because it transfers the message from DNA to protein (Brenner, Jacob, and Meselson 1958). Many years later Yoshiaki Suzuki isolated the first messenger RNA, the mRNA that encodes the silk fibroin protein, which is particularly abundant in the silk gland of the silkworm. There is a different

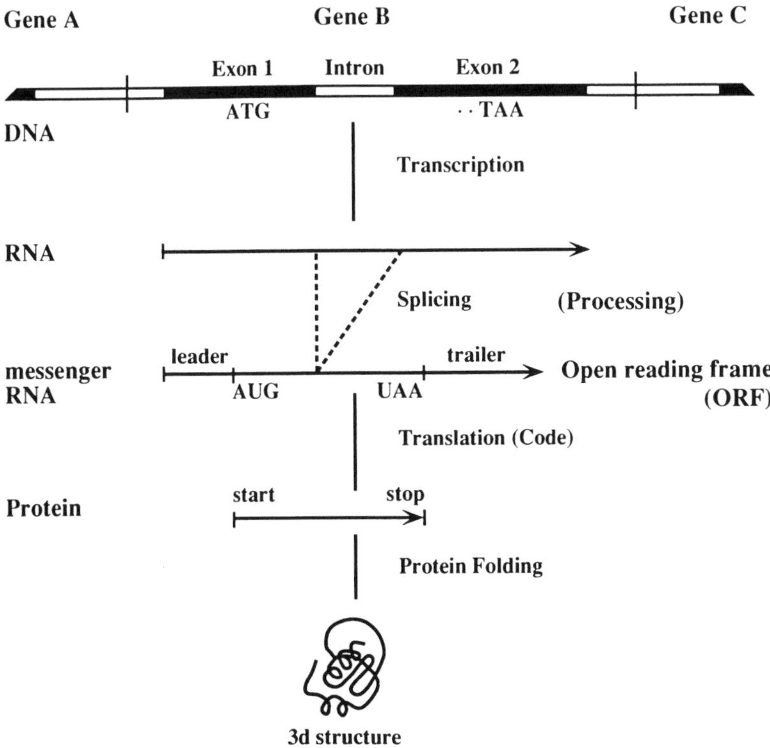

Figure 1.4
Information transfer from gene to messenger RNA to protein

messenger RNA molecule for each protein or set of proteins. RNA molecules are also chains consisting of bases lined up on a sugar-phosphate backbone, but the sugar in this case is ribose rather than deoxyribose, and among the bases thymine (T) is replaced by uracil (U), whereas A, G, and C occur in both RNA and DNA. RNA is also synthesized by an enzyme called RNA polymerase, which is capable of reading stretches of a DNA strand corresponding to one or a group of genes and copying it into RNA (fig. 1.4). This process is called transcription. In eukaryotic organisms, which in contrast to bacteria have a true cell nucleus, DNA replication and RNA transcription occur in the nucleus, whereas protein synthesis takes place in the cytoplasm in small particles called ribosomes. Ribosomes serve as translators: they translate the message contained in the DNA from the language of nucleic acids (DNA and RNA), which comprises four letters in its alphabet, to the language of proteins, with a twenty-letter alphabet. Ribosomes are somewhat like tape recorders: they use the messenger RNA as a tape, decipher the information on the message, and translate it into an amino acid sequence constituting the respective protein.

The genetic code was cracked in 1966. The breakthrough came with a famous experiment of Marshall Nierenberg and Heinrich Matthaei, who originally designed it as a control experiment. Ribosomes were provided with an extremely monotonous message, a synthetic RNA consisting only of uracil residues, so-called polyU. As in many experiments, the control was more important than the actual experiment, which consisted of translating a natural RNA. The ribosomes perfectly "understood" the monotonous message and translated polyU into polyPhe, a protein chain consisting exclusively of phenylalanine (Phe) residues. From this experiment the first code word was deciphered. Previous theoretical considerations had led to the conclusion that a code word (codon) had to consist of at least three bases in order to cover all twenty amino acids. With two bases there are only 4^2 (=16) possible combinations, whereas three bases allow for 4^3 (=64) possibilities, which are more than the minimum of twenty needed to encode all amino acids. Furthermore, the analysis of natural amino acid sequences in different proteins and of frame shift mutations had indicated that the code must be based on the figure 3 or a multiple thereof and that the codons do not overlap. Frame shift mutations are caused by the deletion (subtraction) or addition of single base pairs, which shift the reading frame of the DNA and lead to a completely different amino acid sequence beyond the site of the mutation. The deletion of one or two base pairs results in two different frame shifts, whereas deletion of three consecutive base pairs simply deletes one amino acid from the protein but does not alter the reading frame. These considerations led to the proposal of a nonoverlapping tripled code, with three letters per code word, and the codons for the twenty amino acids were deciphered by using various RNA polymers that could be synthesized at that time. The genetic code (fig. 1.5) can be used as a dictionary to translate nucleic acid sequences into protein sequences. For most amino acids there is more than one codon (see fig. 1.5). The protein chains are terminated by special stop codons that were deduced from chain termination mutants. These mutations lead to premature termination of protein synthesis and shortening of the respective protein chain. The deciphering of the genetic code is a milestone in the history of biology.

The genetic code could have been deciphered according to the same principle that was used by Egyptologist Jean-François Champollion to decode

Aminoacids	Abbreviations		Codewords (Nucleotide triplets)					
Alanine	Ala	A	GCA	GCG	GCT	GCC		
Arginine	Arg	R	CGA	CGG	CGT	CGC	AGA	AGG
Asparagine	Asn	N	AAT	AAC				
Aspartic acid	Asp	D	GAT	GAC				
Cysteine	Cys	C	TGT	TGC				
Glutamine	Gln	Q	CAA	CAG				
Glutamic acid	Glu	E	GAA	GAG				
Glycine	Gly	G	GGA	GGG	GGT	GGC		
Histidine	His	H	CAT	CAC				
Isoleucine	Ile	I	ATA	ATT	ATC			
Leucine	Leu	L	CTA	CTG	CTT	CTC	TTA	TTG
Lysine	Lys	K	AAA	AAG				
Methionine (Start)	Met	M	ATG					
Phenylalanine	Phe	F	TTT	TTC				
Proline	Pro	P	CCA	CCG	CCT	CCC		
Serine	Ser	S	TCA	TCG	TCT	TCC	AGT	AGC
Threonine	Thr	T	ACA	ACG	ACT	ACC		
Tryptophan	Trp	W	TGG					
Tyrosine	Tyr	Y	TAT	TAC				
Valine	Val	V	GTA	GTG	GTT	GTC		
Stop	-	-	TAA	TAG	TGA			

Figure 1.5
The genetic code

the hieroglyphs. Champollion deciphered the hieroglyphs using the Rosetta stone, on which a text written in hieroglyphs was accompanied by its translation into Greek letters. By determining the amino acid sequence of a given protein and the nucleotide sequence of the corresponding DNA, the genetic code can also be deduced, but in the sixties DNA sequencing methods did not yet exist. Today the genetic code has been confirmed many times and by sequencing proteins and DNA from hundreds of species, ranging from bacteria to humans, we now know that the code is universal. The few exceptions that have been reported, in which a few codons are used differently, confirm the universality rule and indicate that the code evolved not by necessity but rather by chance in the course of the history of life. This argues strongly that all existing organisms on earth share common ancestors and are genetically related. This does not necessarily mean that life originated only once; there may have been numerous attempts, but only one was successful in the long run.

One of the major difficulties in explaining the origin of life comes from the finding that, on one hand, DNA requires enzyme proteins for its own replication and the expression of genetic information and, on the other, enzyme proteins are encoded by the DNA. In other words, DNA contains the genetic information, whereas proteins serve a catalytic function, and it is very difficult to explain how these two types of molecules could have evolved simultaneously because they are interdependent; DNA requires proteins and proteins depend on DNA.

Thomas Cech and Sidney Altman have partially solved this puzzle by demonstrating that RNA, which serves as the genetic material in some viruses, can fulfill both purposes. It can store genetic information and also serve some catalytic functions, which, however, are much more limited than those of proteins. Finding catalytic properties in RNA molecules may suggest that RNA preceded DNA early in evolution as the genetic material; the sep-

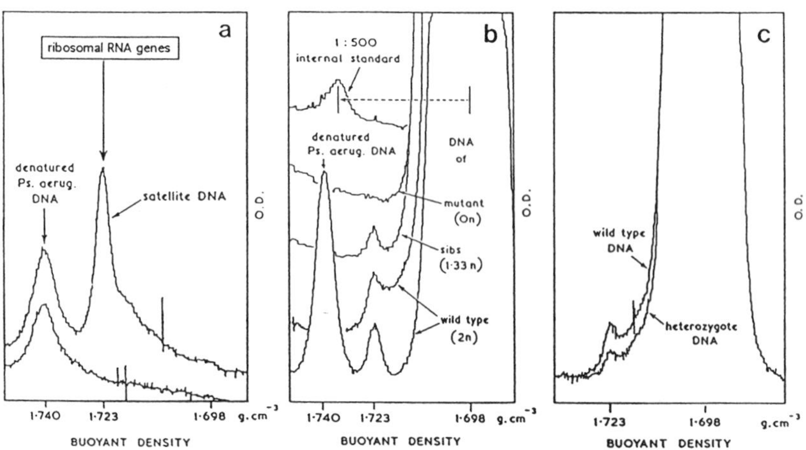

Figure 1.6
Isolation of the ribosomal RNA genes of the frog Xenopus laevis *by density centrifugation: (a) isolation of ribosomal RNA as a peak in the optical density (O.D.) profile at a density of 1.723 gcm⁻³; (b) absence of the peak in the anucleolate mutant lacking ribosomal RNA genes (on), and increase in peak height with increasing gene dosage (1,33n, 2n); (c) 50 percent reduction in heterozygote (1.0n) versus wild type (2n) DNA. From M. Birnstiel, National Cancer Institute Monograph 23 (1996): 431–47.*

aration of information storage and catalytic function into more suitable molecules, that is, DNA and protein, may have occurred only later, with RNA retaining the go-between function as a messenger. RNA may also have retained a catalytic function in protein synthesis, since ribosomal RNA may serve as an enzyme in the formation of the peptide bond between the amino acids during elongation of the protein chains.

Future research on the functions of RNA may provide deeper insight into the problem of the origin of life, which eventually may become amenable to experimentation. However, one of the difficulties with RNA as the original genetic material is its inherent instability. The universality of the genetic code also applies to RNA genomes, which may be witnesses to the ancient RNA world. Yet we cannot rule out the possibility that these viruses originated from DNA genomes. In any case, the fact that the genetic code is shared by all organisms ranging from bacteria to humans argues that the language of the genes is very ancient and may go back to the origin of life at least three billion years ago.

The definitive proof of the chemical nature of the gene was provided by the first isolation of genes by Max Birnstiel and by the total synthesis of a functional gene by Gobhind Khorana and his collaborators. The first announcement of a successful gene isolation was made in 1965 at a meeting in Montevideo, Uruguay, and the news spread like a fire through the scientific world (fig. 1.6). At that time, the mere thought of isolating a particular gene was absolutely amazing. Regrettably, Friedrich Miescher did not live long enough to see his ideas confirmed. Max Birnstiel and his graduate student Hugh Wallace had fractionated DNA from the frog *Xenopus* and succeeded in separating the ribosomal genes from the rest of the frog's genome. These genes encode two major RNA components of the ribosomes (18s and 28s ribosomal RNA), which represent the most abundant stable RNA in cells and therefore can be purified relatively easily. From the base composition of this RNA, Birnstiel had deduced that the genes encoding ribosomal RNA should have a significantly higher density than the average *Xenopus* gene. This allowed him to purify the ribosomal RNA genes by density centrifugation. The isolation of the ribosomal RNA genes was facilitated by the fact that there are several hundred copies of these genes per genome in a tandem-like arrangement, which also came as a surprise. These findings were corroborated by the *anucleolate* mutant, which served as a control, since it lacks ribosomal RNA genes. The experiments clearly show that the satellite DNA peak containing the ribosomal RNA genes is missing in the *anucleolate* (on)

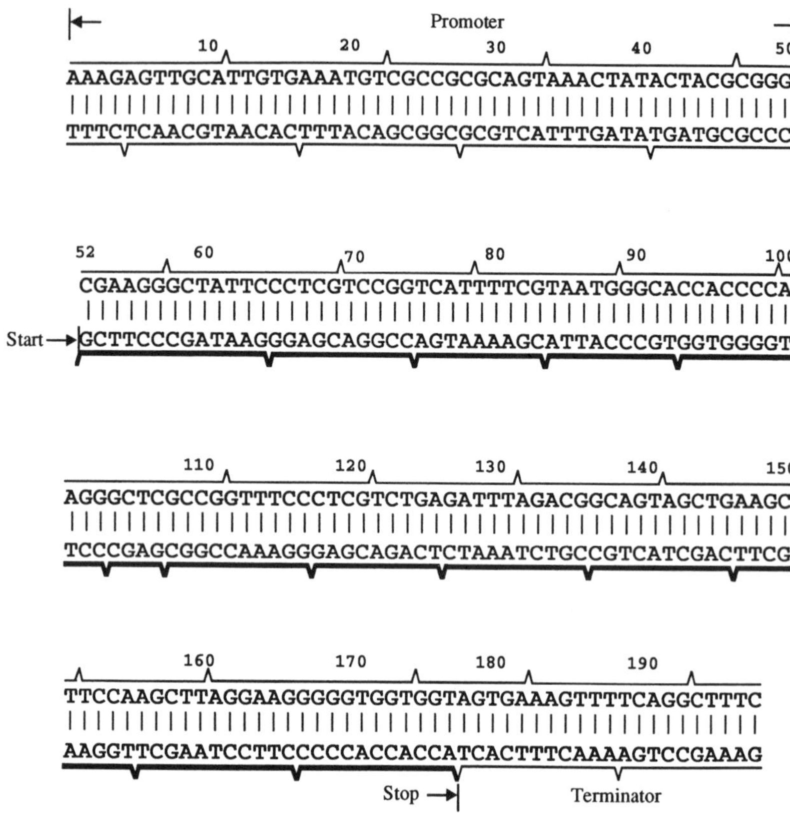

Figure 1.7
Total synthesis of a gene. An active tyrosine suppressor transfer RNA gene was synthesized by joining single nucleotides to form oligonucleotide chains 10–12 units long, which were then joined (at the sites indicated by brackets) to constitute a complete gene of 200 base pairs. The coding region is preceded by a promoter and followed by a terminator region where the synthesis of the gene product (transfer RNA) starts and terminates, respectively. In living bacterial cells this gene is transcribed into functional transfer RNA molecules by the RNA polymerase enzyme. From H. G. Khorana, in T. Sekiya et al., Total synthesis of a tyrosine suppressor transfer RNA gene, Journal of Biological Chemistry *254 (1979): 5787–801.*

mutant, whereas heterozygous animals lacking half of the genes contain intermediate levels (see fig. 1.6). With the advent of recombinant DNA technology, gene isolation eventually became a routine procedure.

The most rigorous proof that DNA is the genetic material was provided by Gobhind Khorana and collaborators, who chemically synthesized a gene from its building blocks, the nucleotides. They first joined the nucleotides into single-stranded fragments of ten to twelve units, annealing the overlapping fragments to form the corresponding duplex molecules, and joining the duplexes to form a DNA molecule two hundred base pairs long (fig. 1.7). This DNA molecule, which encodes a transfer RNA (tyrosine suppressor transfer RNA), was introduced into bacteria and bacteriophages and shown to be biologically active. This first total synthesis of a gene represents another milestone in the history of molecular biology.

How is genetic text read? How is genetic information expressed? This is a particularly important question in higher organisms in which cells differentiate and serve different functions. Why do only red blood cells produce hemoglobin, and why is insulin synthesized only by special cells of the pancreas? The famous German zoologist August Weismann proposed that only the germ cells have a complete genome and therefore that they are totipotent and can give rise to a complete organism. The cells of the body, the somatic cells, were supposed to contain only those genes (which he called determinants) required for their functioning. According to Weismann, genes were segregated in the course of cell divisions and channeled into those cells, in which they are required, and they were lost from the other cells. This hypothesis, however, was difficult to reconcile with the fact that many organisms are capable of regenerating missing parts, indicating no irreversible loss of genetic information from the somatic cells. Weismann could have argued that a small population of somatic cells remained totipotent and formed the source of the regenerated parts. That all somatic cells with few exceptions contain a complete (diploid) set of chromosomes provided no support for Weismann's hypothesis either, and there was no satisfactory explanation for cell differentiation. The idea that the somatic cells have a complete genome but that only a fraction of their genes is actively expressed in a precise spatial and temporal pattern was first formulated by Thomas Hunt Morgan in 1934:

> The common meeting point of embryology and genetics is found in the relation between the hereditary units in the chromosomes, the genes, and the protoplasm of the cell where the influence of the genes

> comes to visible expression. . . . The implication in most genetic interpretation is that all the genes are acting all the time in the same way. This would leave unexplained why some cells of the embryo develop in one way, some in another, if the genes are the only agents in the results. An alternative view would be to assume that different batteries of genes come into action as development proceeds.

And further,

> The idea that different sets of genes come into action at different times is exposed to serious criticism, unless some reason can be given for the time relation of their unfolding. The following suggestion may meet these objections. It is known that the protoplasm of different parts of the egg is somewhat different, and that the differences become more conspicuous as the cleavage proceeds (i.e. the egg undergoes cell divisions), owing to the movements of materials that then take place. From the protoplasm are derived the materials for the growth of the chromatin and for the substances manufactured by the genes. The initial differences in the protoplasmic regions may be supposed to affect the activity of the genes. The genes will then in turn affect the protoplasm, which will start a new series of reciprocal interactions. In this way we can picture to ourselves the gradual elaboration and differentiation of the various regions of the embryo.

This theory of differential gene activity controlling development proved to be essentially correct, but the evidence supporting it was accumulated very slowly.

The first breakthrough came from the work of François Jacob and Jacques Monod, not on multicellular organisms, but on bacteria and bacteriophages, which are much more readily amenable to molecular genetic analysis. As Jacob described vividly in *The Statue Within,* in the 1950s the Pasteur Institute was a mecca for molecular biologists. Specifically, Jacob, Monod, and their colleagues studied adaptive enzymes in *Escherichia coli,* a common bacterium that colonizes the human gut and is normally quite harmless. When grown on glucose, this bacterium does not produce the enzymes required to metabolize lactose, another sugar, even though it has the genes encoding these enzymes. These genes are thus inactive or repressed. If, however, lactose is provided in the medium, the enzymes are rapidly produced. Therefore, gene activity is differentially regulated in response to an environmental stimulus, the sugar. In a series of beautifully elegant experiments, Jacob and Monod showed that the enzymes involved in lactose fermentation are encoded by a closely linked battery of genes constituting a functional unit called the *lac* operon. A separate gene called the repressor gene encodes a repressor, later shown to be a protein, which in the absence of lactose binds to specific DNA sequences in the *lac* operon and prevents transcription. Lactose serves as an inducer of gene activity because it can bind to the repressor that alters its structure, so that the repressor falls off the DNA and the operon can be transcribed. This type of regulation of gene activity is called negative because it involves repression of the operon and de-repression in response to the inducer. Other cases were later found in which regulation is positive and involves synthesis of an activator of gene activity. Bacteria therefore have special regulatory genes whose sole function is to regulate the activity of their target genes. Based on their experiments using bacteria, Jacob and Monod proposed models for how cell differentiation might be controlled in higher organisms, but it took considerable time and effort to identify the regulatory genes that control cell differentiation and development in higher organisms.

Figure 2.1
The Buddha of Nara and the eight-legged butterfly. Photographs by the author.

~

2
The Key to Understanding Development

Homeotic Genes

WE WERE entering a large wooden building, as Tetsuja Ohtaki and his wife guided me through Nara, the old capital of imperial Japan, to visit the Daibutsu, the beautiful statue of the Buddha, and to introduce me to the treasures of Japanese culture. At the entrance, the large wooden house was guarded by two ferocious-looking giant statues, and after passing through the courtyard, we entered a vast, scarcely lit room and were standing in front of the Daibutsu, the gigantic statue of the Buddha. This bronze statue is more than sixty feet high and shows the Buddha sitting with one hand raised peacefully. I shall never forget the incredibly beautiful expression of his face. In front of the Buddha are bronze sculptures of lotus flowers and four large bronze butterflies that attracted my attention because I have loved butterflies since childhood. After closer examination of the two butterflies, I discovered that they had eight legs instead of the six legs characteristic of insects (fig. 2.1). These butterflies clearly represent homeotic mutants in which the first abdominal segment is transformed into an additional thoracic segment with legs. To my knowledge this sculpture, dating from the ninth century, is the earliest representation of a homeotic mutation. In his wisdom, the Buddha tried to show us where the key to understanding development lies, but we did not listen for more than a thousand years. My Japanese colleagues did not know about this treasure, and later I was told that my discovery made the headlines in Japanese newspapers.

What are homeotic mutants? The term *homeosis** goes back to William Bateson's book *Materials for the Study of Variation, Treated with Especial Regard to Discontinuities in the Origin of Species,* published in 1894. Thirty years after the publication of Darwin's *Origin of Species,* Bateson realized that many of the fundamental questions about evolution had not been solved. Darwin's hypothesis implied that evolution was based on random variation and subsequent selection of the fittest, but the nature and the causes of these variations remained obscure. Even though Darwin clearly saw that these variations had to be heritable in order to contribute to evolution, he did not distinguish between heritable and acquired characters and their variations. In an attempt to understand the nature of these variations, which lead to the evolution of new species, Bateson devoted his six-hundred-page book to a description of these variations, especially discontinuous variations because all species were thought to be discontinuous. He subdivided variations into meristic and substantive variations. Numerical or geometric changes he designated as meristic variations; for example, the flower of a *Narcissus* is commonly divided into six parts, but through meristic variation it may be subdivided into seven parts or into only four. Bateson thought that meristic variations were caused physically, because the repeated parts of the flower arise by a process of division from undifferentiated tissue. The second kind of variation was designated as substantive and includes, for example, color variation of the narcissus flower or of the human eye. These variations were thought to be largely chemical in nature.

Even though Bateson realized that variation had to occur during the succession from parent to offspring in order to affect evolution, he completely neglected the fact that variations must be heritable in order to have evolutionary consequences. He consciously avoided the use of the terms *heredity* and *inheritance* and only described the variations in an exhaustive catalog. In describing meristic variations, Bateson recognized that changes in numbers

*This word, also spelled *homoeosis,* derives from the Greek *homoiosis,* assimilation, resemblance. However, when in 1969 I polled my colleagues at Yale Medical School, not one person could read the original Greek spelling. And so I decided to modernize the spelling, replacing *oi* by *e.* Some purists continue to champion the original spelling, but this new spelling has now been adopted by the major scientific journals. After all, the human language evolves like the language of the genes!

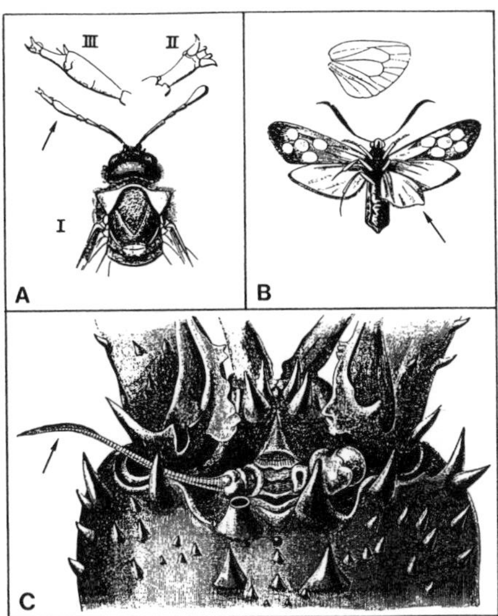

Figure 2.2
Homeosis. Homeotic transformations (arrows) of antennae into legs (A), hind legs into hind wings (B), and eye into antenna (C) in insects and crustaceans. From William Bateson, Materials for the Study of Variation *(New York: Macmillan, 1894).*

can be brought about in two ways, by addition or subtraction, as for example in the number of legs or body segments in *Peripatus,* a primitive animal with twenty-nine to thirty-four segments. Alternatively, one structural element may be transformed into another element; for example, the modification of the antennae of an insect into a foot or the eye of a crustacean into an antenna (fig. 2.2). This phenomenon had been described in plants by Goethe as "Metamorphose," and Masters in his treatise on "Vegetable Teratology" borrowed this term and called it "metamorphy," but this designation is confusing because metamorphosis has a different meaning in developmental biology. Therefore, Bateson proposed a new term, *homeosis,* and defined it loosely as a change of something into the likeness of something else.

From the large collection of homeotic variations described by Bateson I shall mention a few striking examples here. In the sawfly *Cimbex* (fig. 2.2A), the antenna is transformed into a foot—that is, a head segment is partially transformed into a structure from a thoracic segment, which reflects a transformation in the anterior-to-posterior direction. The *Zygaena* moth (fig. 2.2B) shows the transformation of a hind leg into a hind wing that corresponds to a ventral-to-dorsal transformation within the same body segment. Even more striking is the transformation of an eye into an antennalike structure in the crab *Palinurus* (fig. 2.2C). However, homeotic variations are not confined to arthropods, they are also found in mammals and humans. For example, an additional rib may be found in certain human individuals on the seventh cervical vertebra, which corresponds to a transformation of the seventh cervical vertebra into a first thoracic vertebra with a rib. There are also individuals with supernumerary nipples and mammae on the front of the trunk. And some homeotic variations affect the body plan even more profoundly; for example, they affect the radial symmetry of sea urchins and starfish. Normal starfish have five arms and are radially symmetrical, even though they develop from larvae that are bilaterally symmetrical. However, Bateson describes variations of six-armed starfish and four-rayed sea urchins. In none of these cases does Bateson attempt to distinguish between hereditary changes—that is, homeotic mutations—and developmental malformations. For example, it is not immediately obvious whether a six-armed starfish is carrying a homeotic mutation or whether the additional arm is caused by a developmental malformation or abnormal regeneration. Starfish can accidentally lose an arm and subsequently regenerate it. But sometimes two arms are regenerated instead of one. The only hint in Bateson's book that the six-armed starfish (*Asterias rubens*) may be a mutant comes from his remark that six-armed starfish are frequent in the coastal waters at Wimereux, France, which may indicate a homeotic mutation that had become established in this local starfish population. The six-armed starfish and its normal companion (fig. 2.3) were collected in Banyuls, southern

Figure 2.3
Normal (five-armed) and variant starfish (Echinaster sepositus) *with six arms*

France, and when I brought them to the Biozentrum in Basel, the five-armed starfish lost one of its arms to cannibalism, so that there were four, five, and six-armed specimens in the same aquarium. Before I had a chance to determine whether the six-armed specimen was a mutant by breeding it, however, someone stole it from the tank. In another species of starfish (*Asterina gibbosa*), it has been shown that four-armed individuals can arise from developmental malformations during metamorphosis.

Bateson immediately recognized the importance of Mendel's work after it had been rediscovered, and in 1913 he published *Mendel's Principles of Heredity*. In this book Bateson describes a case of brachydactylism somewhat different from the one I mentioned in Chapter 1. He considers the variation caused by this dominant mutation to be homeotic because in this case all the fingers are not only shorter but transformed "into the likeness of the thumb." He also returns to homeotic variation when discussing the transformation of stamens into petals, which occurs for example in roses and ranunculus, or the transformation of "sepals into the likeness of the petals." We shall see later that in at least some homeotic mutants the transformations are actually complete, not merely into "the likeness of some other structure."

Definitive evidence for a genetic basis of some homeotic transformations was obtained by isolating homeotic mutants. The first homeotic mutant was found by Calvin Bridges in Thomas Hunt Morgan's laboratory, the famous fly room, in 1915. This mutant shows a partial duplication of the thorax and was therefore named *bithorax* (*bx*). It transforms the third thoracic segment (T3) (fig. 2.4) toward the second (T2). *bithorax* arose spontaneously in the laboratory and has been maintained continuously as a laboratory stock ever since.

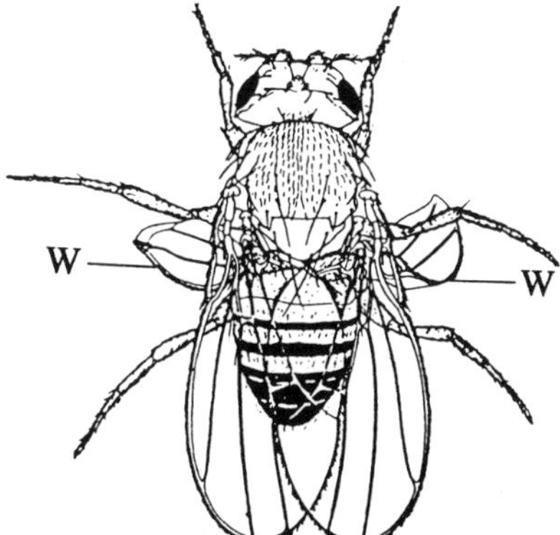

Figure 2.4
The Bithorax mutation discovered by Calvin Bridges in 1915. The balancer (halteres) of this mutant are transformed into winglets (w). From T. H. Morgan, C. B. Bridges, and A. H. Sturtevant, The Genetics of Drosophila *(New York: Garland, 1988).*

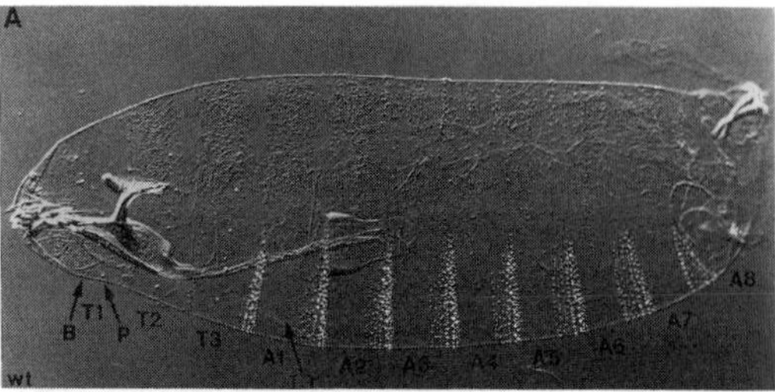

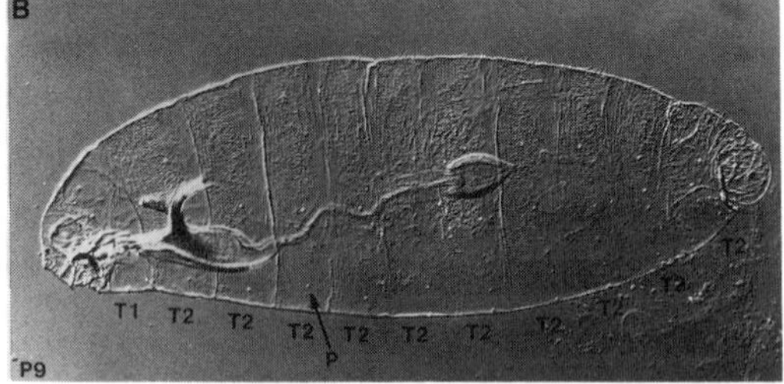

Figure 2.5
The Bithorax complex. Lack of all the genes of the Bithorax complex (deficiency P9 in B) leads to the transformation of all posterior body segments to mesothorax (T2), as compared to the wild-type embryo shown in A. The thoracic segments are characterized by different denticle belts and ventral pits (P) that are absent in the abdominal segments. The prothoracic segment (T1) carries a beardlike structure (B). Courtesy of U. Kloter.

This mutation became the starting point for an extensive genetic analysis of the *bithorax complex* by Edward B. Lewis, who has devoted most of his life to studying these genes. The genetic analysis revealed that the *bithorax complex* consisted of a number of closely linked genes that supposedly arose by tandem duplication in the course of evolution (see Chapter 12) and subsequent divergence in function caused by mutations. A deletion of the entire *bithorax complex*—removing all the genes—is lethal, but the embryos develop far enough to be analyzed morphologically for their segmentation pattern; in these lethal embryos all abdominal segments are transformed into second thoracic segments, which Lewis considers to be the ground state (fig. 2.5). To assess the normal, or wild type, functions of the *bithorax* genes, Lewis constructed a set of genotypes in which the entire *bithorax complex* was lacking, except for the presence of one dose of each consecutive gene. In this way, he could show that all of the genes (or chromosomal regions) of the *bithorax complex* are involved in inducing, or in some cases suppressing, the formation of specific organs and structures like legs or sensory organs. This analysis of the function of the wild type genes of the *bithorax complex* revealed several basic properties of the system. First, there seemed to be one gene (or chromosome region) primarily required for each segment. Second, the genes are arranged in the chromosome in the same order in which they are expressed along the anteroposterior axis of the organism, as assessed by their effects on morphology. This colinearity rule turns out to apply to other organisms as well. Third, the morphological evidence showed that a gene primarily expressed in a given segment tends also to be expressed in the more posterior segments. Therefore, beginning with the ground state, one additional gene is expressed consecutively in each of the more posterior segments until all genes are expressed in the last abdominal segment (fig. 2.6). Consistent with this interpretation, Lewis found two basically different types of mutants that have opposite effects and were designated as loss-of-function versus gain-of-function mutants. Loss-of-function mutants inactivate, or in the extreme case delete, the gene entirely and lead to transformations toward the thoracic ground state—that is, in the anterior direction. To the contrary, gain-of-function mutants show transformations in the opposite direction, away from the ground state. The nature of these gain-of-function mutants, which are dominant over the wild type gene, remained enigmatic until the advent of molecular genetics.

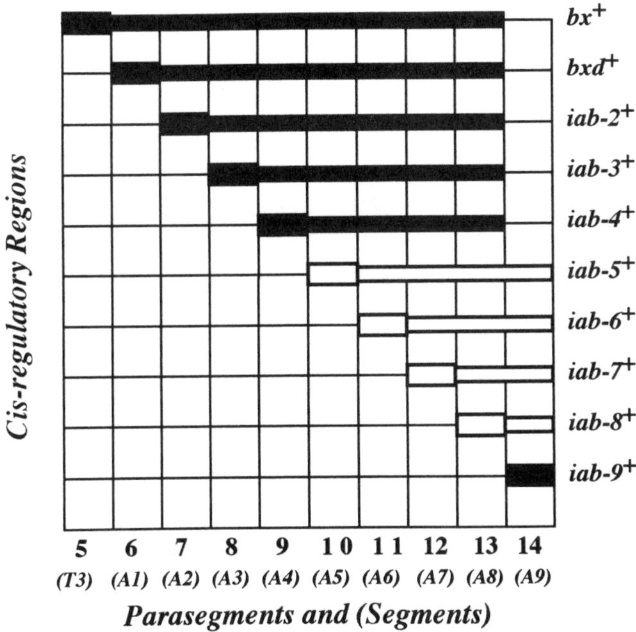

Figure 2.6
The Lewis model. Specification of the posterior parasegments (or segments) by the genes of the Bithorax complex or their respective cis-regulatory regions: Ultrabithorax (bx$^+$ *to* iab4$^+$), abdominal-A (iab5$^+$ *to* iab8$^+$) *and* Abdominal-B (iab-9$^+$). *From E. B. Lewis, Clusters of master control genes regulate the development of higher organisms,* Journal of the American Medical Association 267 (1992): 1524–31.

Figure 2.7
The normal and four-winged fly. Courtesy of E. B. Lewis.

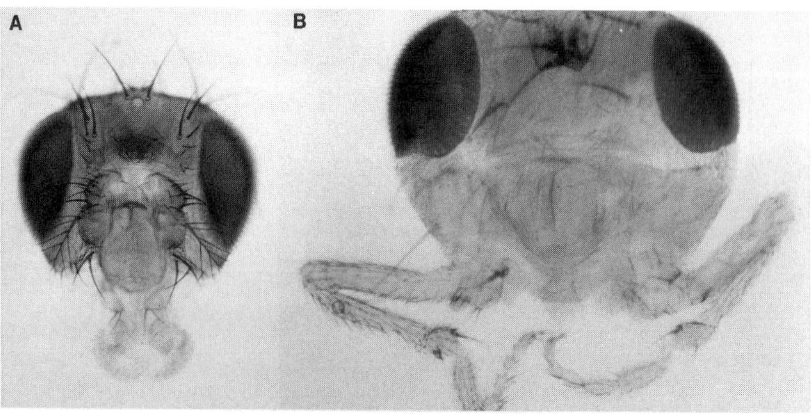

Figure 2.8
Nasobemia. Head of a wild-type (normal) fruit fly as compared to an Antennapedia *mutation* ($\text{Antp}^{Ns}/\text{Antp}^{73b}$)*, which shows a transformation of the antennae into a pair of middle legs.*

The "Lewis Model," as it emerged from this work, is based on genetic, developmental, and evolutionary considerations. It assumes that the strategy for achieving complexity by duplicating and subsequent diversifying is used at both the genetic and the developmental level. Early in embryonic life insects and most higher organisms consist of little more than a tandem array of duplicated body segments. During the course of development, these segments diverge and diversify to produce a variety of segment-specific structures. This strategy is also reflected in evolution; insects most likely evolved from ancestors that consisted of a large number of morphologically similar body segments, each carrying a pair of legs. Insects diversified these segments by diversifying their homeotic genes. The genes of the *bithorax complex*, for example, remove the legs from the abdominal segments, leaving only three pairs of legs on the three thoracic segments and giving each segment its identity. By inactivation of the gene specifying the first abdominal segment, Lewis constructed the eight-legged fly, which corresponds exactly to the eight-legged butterfly of the Buddha mentioned in Chapter 1.

Another spectacular genetic construction is the four-winged fruit fly (fig. 2.7). *Drosophila* belongs to the Diptera, the order of insects with two wings only, whereas most other insects have four wings. In diptera, the hind wings on the third thoracic segment (T3) are reduced to small balancers, called halteres. Inactivation of the *Ultrabithorax* (*Ubx*) genes in the third thoracic segment leads to the transformation of T3 toward T2 and to the conversion of halteres into a second pair of wings. Both the eight-legged and the four-winged fly represent reversions toward the ground state and presumably correspond to an earlier evolutionary stage. The evolutionary wheel of time, therefore, can be turned back. However, the four-winged fly has no muscles in the duplicated thoracic segment and is therefore a poor flier. It will require more genetic work to construct the perfect tetrapteran fly. The opposite, the fly with four halteres, can be constructed by using the dominant gain-of-function mutations *Contrabithorax* (*Cbx*) or *Haltere mimic* (*Hm*). This fly resembles the wingless females of certain other insects in which only the males can fly. These genetic constructs suggest that homeotic genes specify the body plan, the "architecture" of the fly.

The strategy for achieving complexity is the same at the genetic level as it is at the developmental and evolutionary levels and is based on the same basic mechanism shared by all living organisms: reproduction. A series of duplication events leads to the generation of new genes, and subsequent mutations allow their diversification. As a consequence of duplication and diversification of the homeotic genes, the respective body segments also diversify. This suggests that the homeotic genes are master control genes in development and evolution. Yet a formal genetic analysis relying on morphological criteria only has its limitations, and it was the advent of recombinant DNA technology that allowed the elucidation of the molecular basis of homeotic gene action.

My interest in homeotic genes goes back to my days as a graduate student in Ernst Hadorn's laboratory. On examining one of the stockkeeper's fly stocks, I discovered that it contained flies with remarkable homeotic transformations of the head; in the extreme, the antennae were transformed into complete middle legs and the adjacent parts of the head were converted to parts of the ventral thorax (pl. 1, fig. 2.8). It looked as if these animals could walk on their heads. And so I named this mutant *Nasobemia* (*Ns*). The "Nasobem" is an imaginary animal based on a satirical poem by the German poet Christian Morgenstern, which reads in Max Knight's translation:

The Nasobame

Upon his noses stalketh
around—the Nasobame;
with him, his offspring walketh.
He is not yet in Brehm.
You find him not in Meyer
nor does him Brookhaus cite.
He stepped forth from my lyre
the first time into light.
Upon his noses stalketh
—I will again proclaim—
(with him his offspring walketh),
since then, the Nasobame.

An entire biological culture has grown around this poem. It inspired Harald Stümpke (alias Gerolf Steiner) to write a book on a previously unknown group of mammals, the Rhinogradentia, or snouters, which he discovered in the Hi-iay-iay Archipelago in a remote area of the Pacific Ocean. Stümpke describes their embryology, anatomy, physiology, and evolution, based largely on the work of Bromeante De Burlas, one of my teachers at the University at Zurich. Stümpke excels in the description of the

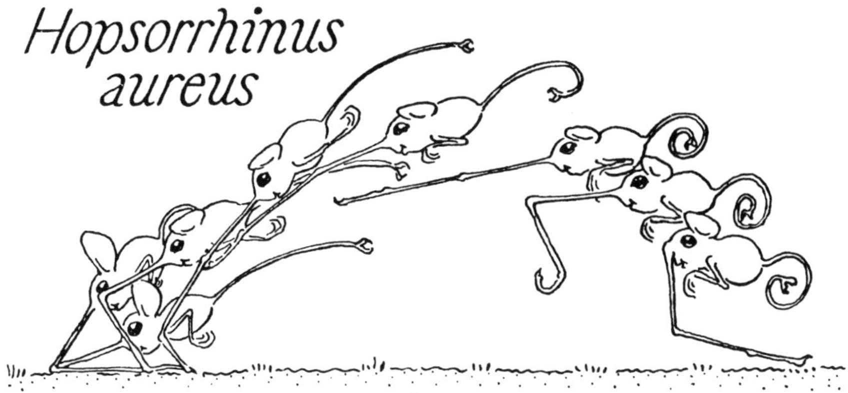

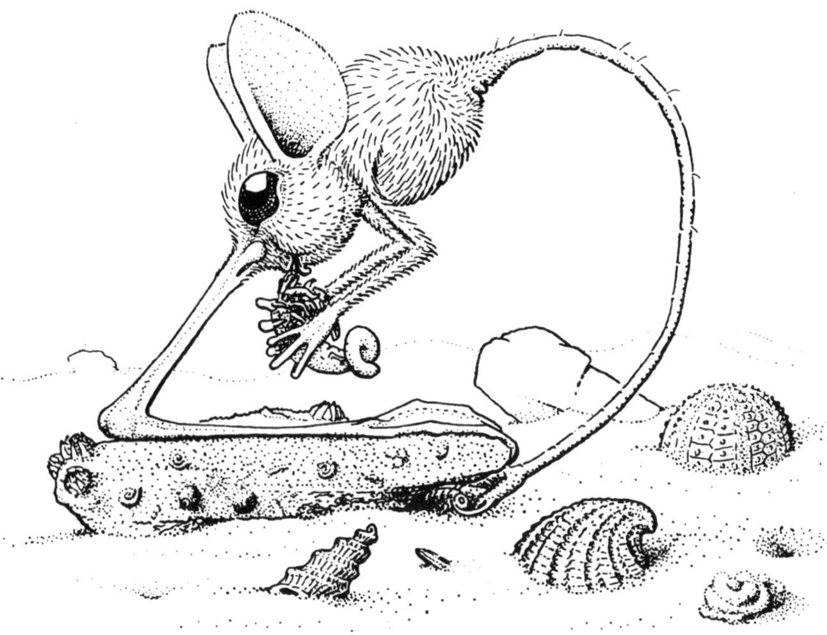

Figure 2.9
The golden nosehopper (Hopsorrhinus aureus). *From H. Stümpke,* Bau und Leben der Rhinogradentia *(Stuttgart: Gustav Fischer Verlag), plate 6.*

various species; I particularly like *Hopsorrhinus aureus*, the golden nosehopper, illustrated here with a time-lapse drawing showing how it moves on its nose (fig. 2.9). He also describes the ferocious predator *Thyrannonasus imperator*, which is rumored to be the hero in Steven Spielberg's next film, so I don't want to say too much more about this fascinating animal. Stümpke makes ample reference to Christian Morgenstern and has named one of the species *Nasobema lyricum* in honor of the great poet. Unfortunately, just as the manuscript on the snouters was ready to go to press, a top secret atomic bomb test destroyed the entire Hi-iay-iay Archipelago, and Stümpke's book is the only remaining record of these remarkable creatures.

The first homeotic mutant causing the transformation of an antenna to a leg was found in 1929 by Elizaveta Balkaschina, who collected a mutant specimen in nature. (It is important to recognize that such mutations are not only generated by geneticists artificially but can arise spontaneously both in the laboratory and in nature.) Balkaschina described her mutant as *aristapedia* because the arista, a featherlike structure on the antenna, is transformed into a foot consisting of four joints with claws at their tip. However, not only do *Nasobemia* flies have a pair of feet at the tips of their antennae, but in extreme cases, they have complete middle legs in place of the antennae. The first mutant of this kind was described by Jean Le Calvez in 1948. It was induced by neutron irradiation and was associated with a chromosome inversion. Inversions are caused by two breaks of the chromosome and reintegration of the intervening chromosome segment in the opposite orientation, which can be visualized in the giant polytene chromosomes (fig. 2.10). Mutations usually occur at one of two chromosomal breakpoints, so that it is not immediately obvious where the mutation is localized. Furthermore, inversions suppress recombination, so that Le Calvez was unable to map his mutation—to assign it to a specific chromosomal location. The two breakpoints were mapped at 84A and 92A, respectively. Because of the mutation's similarity to Balkaschina's mutant, Le Calvez thought that it was a dominant allele (variant) of *aristapedia,* which proved incorrect. For many years Le Calvez's mutation was later thought to have been lost and for many years was not seen, but Rick Garber, while working as a postdoctoral fellow in my laboratory, "rediscovered" the mutant in a stock collection and identified it on the basis of the chromosome inversion. The stock had been mislabeled, and another mutation had been lost instead.

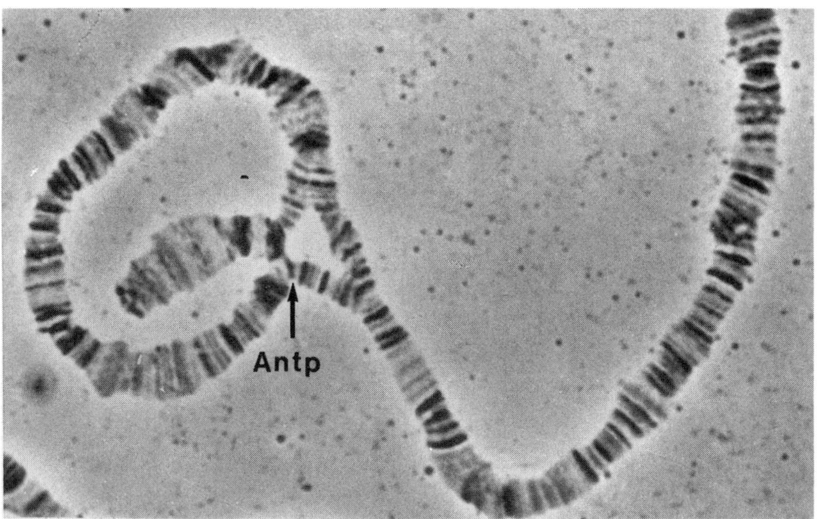

Figure 2.10
Localization of the Antennapedia *gene in the giant chromosomes. The* Antennapedia *gene (*Antp*), discovered by Jean Le Calvez in 1948, is located at the proximal breakpoint (arrow) of the inversion (In(3R)*$Antp^{LC}$*). Courtesy of R. L. Garber.*

In the same year as Le Calvez made his discovery, 1948, Sien-chiue Yu, a graduate student of Ed Lewis, used X-rays to induce a similar dominant mutation in which the antenna is transformed into a middle leg, but with a recognizable arista usually present, unlike *aristapedia,* in which the arista becomes tarsuslike. Yu suggested the name *Antennapedia* (*Antp*) for this mutant type. His mutation was also associated with a chromosomal rearrangement, a translocation with in this case four breakpoints (at 22B, 38F, 83E/F, and 98A). The 83E/F breakpoint is near 84A, one of the breakpoints found by Le Calvez. Subsequently, another four mutations causing the transformation of the antennae into middle legs were found to be associated with inversions having one breakpoint in the 84A area. The chromosomal location of *Antennapedia* thus clearly differs from that of *aristapedia,* which corresponds to section 89C.

The *Nasobemia* mutant is not associated with any visible chromosome rearrangement, which allowed me to map it by recombination. Because few recombinants were found between *pink* (*p*), an eye-color mutant, and *Nasobemia,* I determined that the two genes must be closely linked and map near each other. This genetic map position corresponds exactly to section 84A in the giant polytene chromosomes, where a strong band of compacted DNA is visible. At the time, I could not decide whether the *Nasobemia* mutation affected the same gene as *Antennapedia,* because there was no decisive test for answering this question for dominant mutations. Yet both mutations also have a recessive effect that manifests itself only in homozygous animals with two mutant copies of the gene. Most *Antennapedia* mutants are homozygous lethal and die either as embryos or as larvae, and homozygous *Nasobemia* mutants are at least semilethal because only 40 percent survive. When I crossed *Nasobemia* with *Antennapedia,* a substantial fraction of the compound flies, carrying one copy of *Antennapedia* and one of *Nasobemia,* survived and showed extremely large and well-developed legs in place of the antennae (see fig. 2.8). The conclusion from this experiment was that either the two mutations are in the same gene or they affect two very closely linked genes with similar function. This question could only be resolved later by cloning the respective genes and through molecular genetic analysis. It turned out that *Nasobemia* is in fact a mutation of the *Antennapedia* gene. Le Calvez's information about the chromosomal breakpoints in 84A and the genetic mapping data were crucial for the later cloning of the gene.*

What fascinated me most about *Nasobemia* was the idea that a single gene can apparently activate all the genes necessary to form a leg. This was obviously a master control gene that decided where on the body legs will be formed. It specified the architecture, the body plan of the fly. I then decided to try to elucidate how such a gene acted, which would provide the key to the understanding of development. In those days, trying to understand the molecular basis of gene action was a formidable task: this was b.c. (before cloning), as my postdoctoral fellows refer to this period today, and on both sides there were numerous critics who tried to discourage me from my plan. Many molecular biologists thought that this system was much too complex ever to be solved in molecular terms, and their reductionist strategy was always to study the simplest possible case. One gene was complicated enough, but a gene that controlled hundreds or thousands of other genes was absolutely hopeless. Others thought that I might spend decades trying to identify the product of homeotic genes out to discover that it was a simple

*In a later cytogenetic analysis by Tom Kaufman the *Antennapedia* gene was finally assigned to 84B rather than 84A.

enzyme that yielded no information whatsoever on development. Classical biologists, in contrast, were convinced that molecular biologists were asking the wrong kinds of questions and believed molecular biology to be a nuisance anyway. Morphogens and positional information were considered to be mystical principles, and the possibility that they might be molecules was dismissed as a materialistic point of view.

Just around the time when I finished my graduate studies, however, the genetic code had been cracked and some of the outstanding molecular biologists were looking for new fundamental biological problems to work on, such as cell differentiation and development and neurobiology and the brain, and some even ventured into behavior. Among them was Alan Garen, who became interested in transdetermination, a phenomenon discovered by Ernst Hadorn and Theo Schläpfer, one of my fellow graduate students. For many years Hadorn had been working on imaginal discs, the primordia of adult structures in *Drosophila*. The *Drosophila* larva contains disc- or saclike structures attached inside to the skin from which during metamorphosis the adult body is built up as from building blocks (see fig. 11.1). Three pairs of leg discs give rise to the legs, a pair of wing discs forms the dorsal thorax and wings, and other discs, such as a pair of haltere discs and a pair of eye-antennal discs, form the respective adult structures of the fly. In the pupal case the individual discs are assembled like a mosaic, each disc contributing a defined part of the adult. The discs are precisely determined—programmed—to form a certain structure, and the individual disc cells, even when intermixed with cells from another disc, differentiate autonomously according to that program. These discs can be transplanted into a host larva whose larval blood serves as a culture medium in which the discs can grow. When the host larva has matured, it secretes the molting hormone ecdysone, which induces pupation and metamorphosis of both the host tissues and the transplanted disc. In the body cavity of the host, the transplanted disc can form a leg or a wing or even a functional eye depending on what kind of disc was transplanted. Hadorn found that one could also skip metamorphosis in transplanting the disc from a larva directly into an adult female fly. Because the blood of the adult flies contains only a low concentration of the molting hormone ecdysone, the transplanted discs grow but do not differentiate into adult structures. By serial transfer, Hadorn was able to culture imaginal disc tissue for up to five years, much longer than a fly ever lived. At each generation, part of the tissue was transplanted back into a larva and induced to undergo metamorphosis. Upon culturing, the tissue derived, for example, from a leg disc, gives rise to leg structures according to their original determination. However, with a certain probability "foreign" structures like wing tissues may also be formed. Hadorn coined the term *transdetermination* to describe this phenomenon. For my thesis project, I cultured antennal discs, and in addition to antennal structures, the cultured disc tissue gave rise to leg, wing, and other foreign structures. These were obviously homeotic transformations, and the antenna-to-leg transformation corresponded to mutations like *aristapedia* and *Antennapedia*. I was subsequently able to show, however, that transdetermination is unlikely to be due to homeotic mutation because it occurs simultaneously in several neighboring cells. Rather, it seems to be induced by interactions among the cells during growth and regeneration. The mechanism of transdetermination is still not understood, but there are indications that it involves changes in the activity of homeotic genes rather than mutations. It seems more closely related to certain regeneration phenomena in stick insects and lobsters. If, for example, the antenna of a stick insect or a lobster is amputated at the base, these animals after several molting periods will regenerate a leg rather than an antenna. Although these phenomena remain unexplained, with appropriate molecular genetic tools they may be solved in the future.

Alan Garen thought that transdetermination might provide the key to understanding cell differentiation, and traveled to Hadorn's laboratory to inquire about the possibility of spending a sabbatical year in Zurich to learn about transdetermination. Hadorn was out of town when Garen arrived, however, so I was put in charge of hosting him and discussing transdetermination with him. At the end of the day, he and I decided that, instead of Alan coming to Zurich, I should go as postdoctoral fellow to Yale. This was a decisive step in my scientific career, one I never regretted, and the beginning of a long-lasting friendship. Yale became my second alma mater; I absorbed all the molecular biology I could, and in return, I taught my adviser whatever I knew about *Drosophila* and transdetermination.

Shortly before I arrived at Yale, Walter Gilbert and Benno Müller-Hill had identified the *lac* repressor as a protein, ending a long debate over whether the final product of the repressor gene was RNA or a protein. It therefore followed that gene activity could be regulated by proteins that bind to specific DNA sequences in those genes that are under their control. Realizing that gene regulation lay at the base of cell differentiation in higher organisms, Alan and I began to purify proteins that bind specifically to DNA, looking for differences among these proteins from different imaginal discs.

The technical difficulties were overwhelming, however, and after some time we had to switch to less ambitious projects. In retrospect it is pleasing to see that we were on the right track, but at the time the methods for finding the needle in the haystack were simply unavailable. We had to wait for the advent of recombinant DNA technology to tackle this challenging problem.

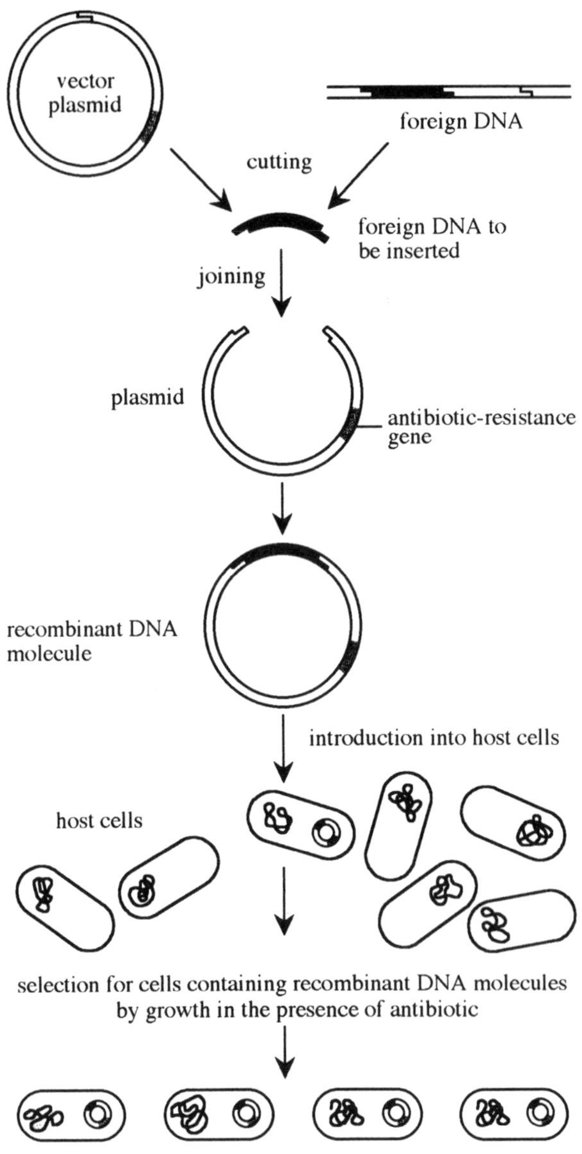

Figure 3.1
Gene cloning. Recombinant DNA molecules generated by cutting and rejoining DNA from a vector plasmid and foreign DNA can be introduced into bacterial host cells. The bacteria can grow into colonies derived from a single cell, or clones. Each clone contains a single recombinant DNA molecule, which it produces in large amounts. From J. D. Watson and J. Tooze, The DNA Story: A Documentary History of Gene Cloning *(San Francisco: W. H. Freeman, 1981).*

~

3
Gene Cloning and the Discovery of the Homeobox

THE ADVENT of recombinant DNA technology revolutionized biology and had serious repercussions on many other aspects of human culture. It opened one of the most fascinating periods in the history of science. A first documentary history of gene cloning was compiled by James Watson and John Tooze in *The DNA Story*. In the present context I can merely outline the concepts and methods used in gene technology so that the uninitiated reader can understand, at least in principle, the applications described in the following chapters.

The various tools for gene cloning gradually accumulated over time until in 1973 it became possible to assemble the pieces of the puzzle and to recombine different DNA molecules in plasmid or viral vectors, introduce them into host cells where they could multiply, and finally re-isolate them in large quantities. This scenario required enzymes for cutting and joining DNA molecules, appropriate vectors (vehicles capable of multiplying in certain host cells), and methods for transferring DNA into these host cells.

Here, briefly, is the procedure for gene cloning in bacterial plasmids (fig. 3.1). Bacteria can harbor, in addition to their single circular chromosome, smaller circular DNA molecules called plasmids, which are easier to manipulate and can be used as cloning vectors. Plasmids contain a DNA sequence called origin of replication that allows them to replicate, or multiply, autonomously. In addition, these plasmids contain an antibiotic-resistance gene, which allows the host cell to grow on a medium containing the antibiotic. Both the foreign DNA and the plasmid vector are cleaved with an enzyme that works like scissors. These enzymes, called restriction enzymes, can be isolated from various bacterial strains, which produce them as part of a defense system directed against foreign intruding DNA, as discovered by Werner Arber. They recognize certain base sequences, such as AATTC, and cut the DNA precisely at these sites. The cut vector DNA molecules can be joined at their ends to a foreign DNA molecule cut by the same enzyme and joined by another enzyme called ligase, resulting in a circular recombinant DNA molecule (see fig. 3.1). This construct can be introduced into host bacteria that are sensitive to the antibiotic and cultured on antibiotic-containing medium. This allows the selection of cells that have taken up a plasmid DNA molecule conferring antibiotic resistance. Because DNA uptake is a rare event, most cells do not receive a plasmid and are killed by the antibiotic, whereas the few survivors usually contain a single plasmid. From the plasmid-containing single cell, a colony of cells can be grown, each cell harboring one or several copies of the same plasmid. Because a family of cells derived from a single cell is called a clone, this procedure has been named gene cloning. The bacterial cells serve as little DNA factories for multiplying a given recombinant DNA molecule, which can then be isolated in pure form from the bacterial culture. Each clone carries a single piece of foreign DNA, so that there is no contamination with other foreign DNA molecules. In addition to plasmids, other vectors, such as bacterial viruses (called bacteriophages), can be used in cloning.

RNA molecules, too, can be cloned. First, a complementary DNA copy (cDNA) must be made from the RNA. This is done using an enzyme called reverse transcriptase, which can transcribe RNA into DNA. The second DNA strand is then synthesized using the cDNA as a template, and the resulting double-stranded DNA molecule can be cloned into a vector as described above.

Cloning is a genetic procedure that allows you to find the needle in the haystack. To purify a DNA molecule a millionfold by ordinary biochemical procedures is extremely difficult, but to select a single cell carrying a recombinant plasmid from a million other cells is easily possible. This illustrates the power of gene cloning.

I remember vividly one of our "journal clubs" held in 1972 in which the recent literature was being discussed in the laboratory, and Elisha van Deusen, one of my former postdoctoral fellows, presented a paper by Paul Berg in which one of the first recombinant viruses was reported. I knew immediately that this was the beginning of a new era in biology. Progress was so rapid and the power of this method was so enormous that scientists became alarmed about potential hazards and called for a moratorium on research. In 1975, at the International Asilomar Conference on Recombinant DNA Molecules, scientists urged the adoption of guidelines regulating recombinant DNA experimentation and called for the development of safe vectors and bacterial strains that could not escape from the laboratory. Guidelines were accordingly established in 1976 and were adopted in different forms by various countries that led in recombinant DNA research, but these rigid regulations turned out to be unnecessary and were relaxed after it became clear that recombinant DNA posed no new dangers.

In the early days of recombinant DNA technology, most of the pioneering work was being carried out in the biochemistry department at Stanford University. The first *Drosophila* genes were cloned by David Hogness and his group at Stanford, and it became clear to me that this was the way to approach the problems of the genetic control of development. I was pleased, therefore, to receive a letter from Paul Schedl from Stanford asking whether he could come as a postdoctoral fellow to my laboratory at the Biozentrum. I accepted him enthusiastically and asked him to acquire the necessary know-how for cloning *Drosophila* genes before coming to Basel, which he did. At around the same time Spyros Artavanis-Tsakonas also joined my laboratory, and he, too, was willing to invest the time to establish a *Drosophila* gene bank at the Biozentrum in Basel. In more civilized places like Stanford they called these collections of recombinant plasmids gene libraries, but in Switzerland the name gene bank seemed more appropriate.

To include the entire *Drosophila* genome with 99 percent probability, one must isolate approximately fifty thousand plasmids with the *Drosophila* DNA inserts having an average length of fifteen thousand base pairs. This was a heroic effort at the time, because we had only three restriction enzymes at hand: one was a gift of Ken Murray in Edinburgh, one was purified in the microbiology department of the Biozentrum, and one was made in my own group. Also, the terminal transferase enzyme, which is required for adding linker tails to the DNA, had to be purified from several pounds of calf thymus, and the laboratory smelled like a butcher shop for a week. We therefore joined forces with Susumu Tonegawa from the Basel Institute of

Immunology, and with the microbiology department's Werner Arber, Bob Juan, and Vincent Pirrotta, and working together we established the first *Drosophila* plasmid library in Basel. The twins, as Paul Schedl and Spyros Artavanis-Tsakonas were called because they collaborated so closely, and Ruth Steward-Silberschmidt isolated almost twenty thousand plasmid clones—enough for a start.

From this gene bank a number of genes, such as the 5s ribosomal RNA genes, were isolated. Genes can be isolated from a gene bank using a method called nucleic acid hybridization. If a solution of double-stranded DNA molecules is heated above its melting temperature, the base pairs, which are held together by hydrogen bonds, split and the two strands separate. When the temperature is lowered again, the single strands can reassociate to form a double helix, but only if the two strands are complementary and in register. If the mixtures of different single-stranded DNA molecules are allowed to reassociate, only molecules that are at least partially complementary can pair and form a double-stranded molecule. One of the partners can also be complementary RNA, which upon reannealing can form a hybrid molecule. There are both RNA-DNA and DNA-DNA hybrids. This hybridization reaction can be monitored by radioactive labeling of one of the strands, for example the RNA. In order to isolate the 5s ribosomal RNA genes, 5s RNA was purified from *Drosophila* ribosomes, labeled radioactively, and hybridized to the entire collection of bacterial clones constituting the gene bank. The labeled 5s RNA hybridized only to the DNA of those bacterial colonies that harbored 5s ribosomal genes of *Drosophila.* The respective colonies were then grown up and the plasmid DNA with the 5s ribosomal RNA genes isolated.

Similarly, the major heat-shock genes of *Drosophila* were isolated in collaboration with Alfred Tissières and his group in Geneva. In response to a brief heat shock or other forms of stress, *Drosophila* cells activate a small number of heat-shock genes, while most of the normally active genes are turned off. The messenger RNAs transcribed from the heat-shock genes are selectively translated into heat-shock proteins, which serve a protective function. To isolate the heat-shock genes, messenger RNA was isolated from heat-shocked *Drosophila* cell cultures in Geneva and shipped to Basel to be hybridized to the gene bank. The labeled heat-shock messenger RNA hybridized only to the DNA from those colonies that harbored the heat-shock genes. These genes hybridized (bound) selectively to heat-shock messenger RNA, which was then translated into specific heat-shock proteins.

Hybridization can also be applied directly to chromosomes, in particular to the giant polytene chromosomes from the salivary glands of *Drosophila*. These chromosomes contain approximately a thousand DNA double helixes that are lined up in parallel so that there are a thousand copies of each gene confined to a narrow space. The DNA strands can be dissociated by heating the chromosome preparation and they can be hybridized to labeled, single-stranded DNA from the cloned heat-shock genes. The labeled DNA will only bind (hybridize) to those chromosomal sites where the heat-shock genes are located. By this procedure, called in situ hybridization, the heat-shock genes can be mapped directly to their respective chromosomal sites. The cloned heat-shock gene can be visualized in the electron microscope by hybridization of the messenger RNA (mRNA) to the hybrid plasmids under conditions in which the RNA-DNA duplex is more stable than the DNA-DNA duplex. Under these conditions the RNA binds to its complementary DNA strand and displaces the second DNA strand, giving rise to a "bubble" (r-loop) representing the cloned gene in the circular DNA molecule (fig. 3.2).

The *white* gene, the first *Drosophila* gene to be identified by mutation, was also isolated by Michael Goldberg and Renato Paro in my laboratory, illustrating the point that genes can be isolated and subsequently reintro-

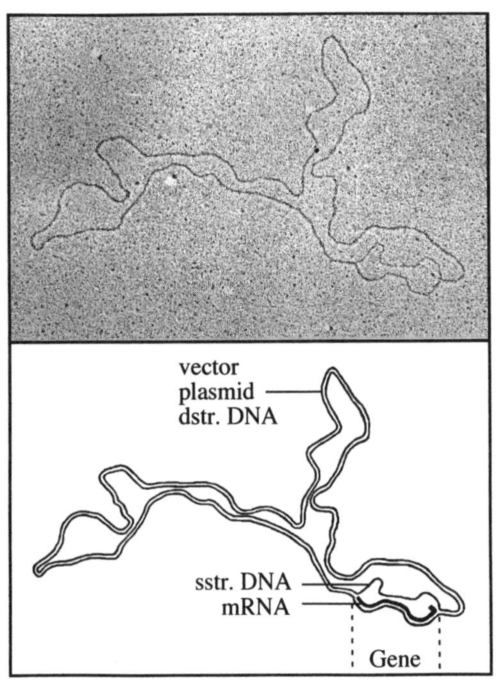

Figure 3.2
Visualization of a gene by electron microscopy. A vector plasmid DNA molecule containing one copy of a heat-shock gene (hsp70) was annealed with its messenger RNA (mRNA) under conditions in which the RNA-DNA hybrid molecule is more stable than the DNA-DNA duplex. Under these conditions the mRNA molecule displaces the corresponding DNA single strand and gives rise to an r-loop (bubble) that represents the gene. Courtesy of J. Meyer.

duced into the germline of *Drosophila* to analyze their function. This method makes use of a transposon, a mobile genetic element, that can jump around in the genome. The P-transposon is a DNA segment of approximately three thousand base pairs encoding a transposase enzyme that catalyzes its own transposition. It is flanked by short, inverted repeat sequences that are recognized by the transposase and allow its integration at more or less any site in the genome. The chromosomal DNA is cut at the target site and the P-element is inserted by the transposase.

Using the P-transposon Allen Spradling and Gerry Rubin developed an efficient method for gene transfer in *Drosophila* (fig. 3.3). The cloned gene to be transferred is inserted as a passenger into a P-vector whose own genes

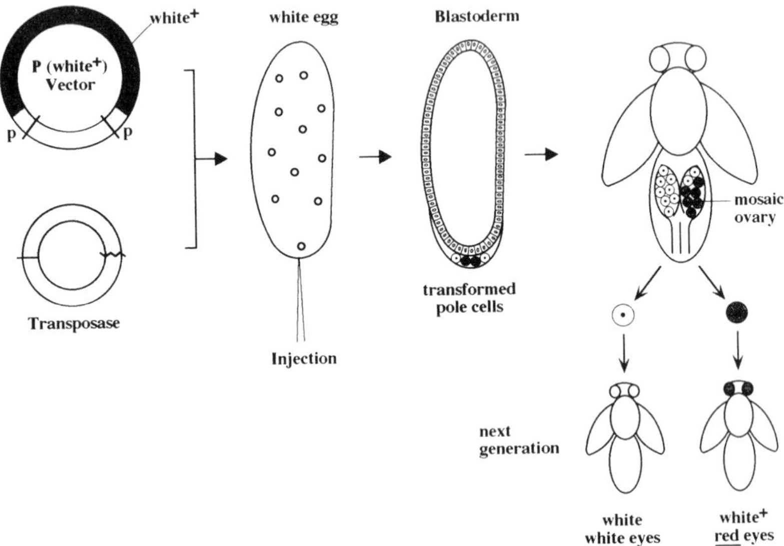

Figure 3.3
Germline transformation in Drosophila. *The P-transposon serves as a vector that can transfer a passenger gene and integrate anywhere in the genome. The* white⁺ *passenger gene, encoding red eye coloration, is inserted into the P-vector DNA and injected with another P-plasmid encoding transposase into fertilized eggs of the* white (w⁻) *mutant recipient (white eye color). The DNA is incorporated into some of the pole cells (future germ cells) of the blastoderm embryo. The transposase gene is activated, and the transposase enzyme causes the P-transposon carrying the* white⁺ *gene to become inserted anywhere into the recipient chromosomes. The embryo develops into a mosaic fly with a mixture of* white⁻ *and* white⁺ *ovarian cells. In the next generation, red eyed* (w⁺) *transformed flies can be obtained in addition to the white eyed progeny of the recipient genotype* (w).

have been removed, leaving only the inverted repeats at both ends. Because such a vector is incapable of transposition, a helper plasmid is constructed to supply the transposase to the vector plasmid. The helper plasmid has defective inverted repeats (its wings have been clipped) so that it is no longer capable of integration into the chromosomes of the host, but it helps the vector plasmid to do so. If the normal *white*$^{+}$ gene is used as a passenger, the two plasmid DNA solutions are co-injected into newly fertilized *white*$^{-}$ (mutant) eggs. At a relatively high frequency, the vector plasmid becomes integrated into the chromosomes of a cleavage nucleus. Cleavage nuclei located at the posterior pole of the egg become pole cells, the future germ cells of the developing fly. The flies that develop from the injected eggs are mated to *white*$^{-}$ partners, and in the next generation progeny derived from a transformed germ cell can be recognized by their red eye color. The site of integration of the P-vector can be mapped in the giant polytene chromosomes by in situ hybridization (fig. 3.4). By such complementation experiments, it can be rigorously proven that the cloned gene is a functional *white*$^{+}$ gene that confers red eye coloration. By mutating the injected gene in the test tube the functional significance of the various DNA segments or the individual base pairs can be analyzed in detail.

The usual procedures for gene isolation require some prior knowledge about the nature of the gene products, but in the case of homeotic genes this information was not available. Methods therefore had to be developed for the cloning of genes for which only some mutants were available with no additional information about their gene products. In other words, all we knew was the genetic and chromosomal map position of the gene. An ingenious method to serve this purpose was developed by David Hogness and his col-

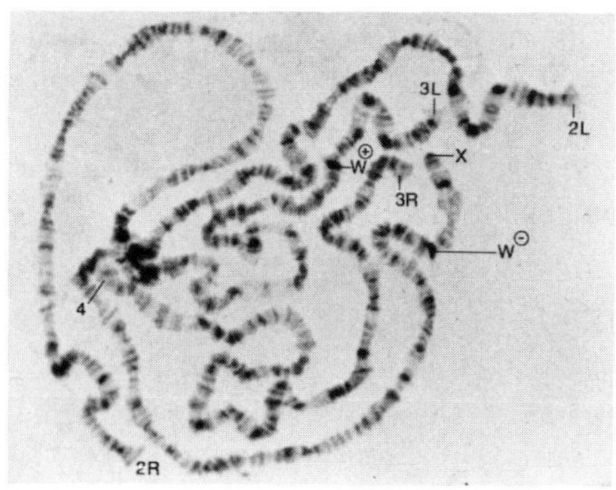

Figure 3.4
Detection of a white^{+}*-transposon in the giant chromosome of a transformed larva. Using the* white^{+} *gene as a probe, the* white^{+} *transposon* (w^{+}) *can be detected on the left arm of the third chromosome, whereas the mutated white gene of the recipient* (w^{-}) *is localized on the X-chromosome. The ends of the chromosomes are labeled* X, 2, 3, *and* 4; L *and* R *stand for left and right arm, respectively. Courtesy of U. Kloter.*

laborators Welcome Bender and Pierre Spierer. This method is called "walking along the chromosome," which sounds like the title of a song from the hit parade, but it involves a lot of hard work. Because the chromosome is a linear structure consisting of a double-stranded DNA molecule, you can "walk" from a previously isolated gene at position A to an unknown gene at position B by isolating partially overlapping DNA segments step by step until you reach B. Partially overlapping DNA segments hybridize to one another, and fragment 1 can be used to isolate fragment 2, 2 can be used to isolate 3, and so on until position B is reached. The direction in which you are walking can be determined, after a sufficient number of steps, by in situ hybridization to giant chromosomes. This method was first used to clone the *bithorax* locus by the Hogness group, and the *Ultrabithorax* gene was the first homeotic gene to be cloned.

As soon as these methods became available, I decided to try to isolate my favorite homeotic gene, *Antennapedia.* Two postdoctoral fellows in my laboratory, Richard Garber and Atsushi Kuroiwa, embarked on a long and strenuous chromosomal walk that lasted more than three and a half years. When we started, we knew only that *Antennapedia* is located in chromosome band 84B, since all chromosome inversions having an *Antennapedia* phenotype with antenna to leg transformations have one of the two breakpoints in this band (see fig. 2.10). At the time we began our walk, however, no cloned sequences from 84B were available. We therefore began walking in band 84F, which is several million base pairs away from *Antennapedia,* and used an inversion called *Humeral* (In(3R)Hu) to "jump" from 84F to 84B. Even so, we had to walk for approximately 250,000 base pairs before reaching the *Humeral* breakpoint in 84F, which allowed us to jump into the *Antennapedia* region. From the *Humeral* breakpoint near *Antennapedia* another 230,000 base pairs had to be cloned, and only thanks to the persistence of Rick Garber and Atsushi Kuroiwa was this task accomplished.

To find out where *Antennapedia* is located on this cloned DNA segment, we first mapped the breakpoints of the various *Antennapedia* inversion mutants, which were located within approximately 50,000 base pairs. Then the gene was roughly defined by chromosomal deletions that either remove *Antennapedia* and leave the flanking genes intact or remove the flanking genes and leave *Antennapedia* intact. This accomplished, we proceeded toward the identification of its messenger RNAs. A cDNA library, made from embryonic *Drosophila* RNA, was kindly provided by Michel Goldschmidt-Clermont, from which several cDNA clones—that is, DNA copies of *Antennapedia* mRNA—were isolated. Even though these mRNAs were only 3,400–5,000 bases long, the *Antennapedia* gene turned out to be extremely large, more than 100,000 base pairs. Like most other genes in higher organisms, *Antennapedia* is a split gene consisting of exons, which contain the coding information, separated by introns (see fig. 1.4). Both exons and introns are transcribed into RNA, but the intron sequences are removed from the primary transcript and the exon sequences are spliced together to form the processed mRNA. *Antennapedia* has extremely long introns and includes large control regions in both its introns and its flanking sequences, which make it one of the largest *Drosophila* genes.

To map the location of the exons, Rick Garber hybridized the cDNA clones to the entire collection of chromosomal clones from the walk. In this experiment only those chromosomal regions that contained exon sequences hybridized with the cDNAs. To his surprise, the *Antennapedia* cDNA also cross-hybridized with sequences outside the *Antennapedia* gene. This was the first sign of the homeobox. It turned out that the *Antennapedia* cDNA hybridized to the neighboring *fushi tarazu* (*ftz*) gene, since the two genes share similar DNA sequences. In fact, I had been looking for such cross-homologies between different homeotic genes, because Ed Lewis had postulated that various homeotic genes had arisen by tandem duplication, which implied that they still share some common DNA sequences. The *fushi tarazu* gene, located next to *Antennapedia*, was first described by Barbara Wakimoto, who gave it its Japanese name, meaning not enough segments, because the mutant embryos lack alternate segments (or parasegments). After Garber's results, I asked my postdoctoral fellow Atsushi Kuroiwa to clone the *fushi tarazu* gene, which I describe in later chapters. It turned out that in both genes, *Antennapedia* and *fushi tarazu*, the cross-hybridizing DNA sequences were confined to the last exon.

Shortly before then, a new collaborator, William McGinnis, had joined my laboratory, and he was enthusiastic about my proposal to define the homologous sequences between *Antennapedia* and *fushi tarazu* more precisely. After working out suitable conditions for low-stringency hybridization, which allows the detection of genes that are similar but not identical to the probe, he scanned the entire genome for sequences that cross-hybridized with the various segments of the *Antennapedia* and *fushi tarazu* cDNAs. He soon found a number of defined DNA fragments that hybridized to those last exon sequences from *Antennapedia* and *fushi tarazu*, representing at least a dozen genes, most of which were later rapidly cloned. To find out whether other homeotic genes also contained this cross-hybridizing DNA sequence, I called Pierre Spierer in Geneva, who had just returned from David Hogness's laboratory, where he and Welcome Bender had cloned the homeotic *Ultrabithorax* gene. I asked Spierer just for two clones covering the first and the last exon of *Ultrabithorax*, which he kindly provided us. Soon, in close collaboration with Michael Levine and Ernst Hafen, Bill McGinnis showed that the last exon of *Ultrabithorax* also cross-hybridizes with these sequences of *Antennapedia* and *fushi tarazu*. Eureka! At that point we knew we had discovered something important, and we coined the name homeobox (fig. 3.5).

Subsequent sequencing (determination of the DNA sequence) ruled out the possibility that the cross-hybridization was due to a simple repeated sequence, and it defined the homeobox as a 180-base-pair DNA segment

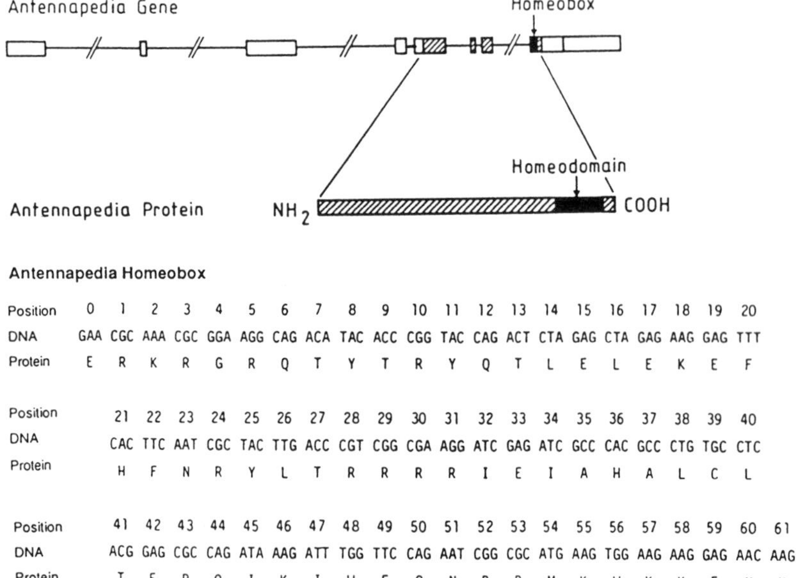

Antennapedia Homeobox

Position	0	1	2	3	4	5	6	7	8	9	10	11	12	13	14	15	16	17	18	19	20
DNA	GAA	CGC	AAA	CGC	GGA	AGG	CAG	ACA	TAC	ACC	CGG	TAC	CAG	ACT	CTA	GAG	CTA	GAG	AAG	GAG	TTT
Protein	E	R	K	R	G	R	Q	T	Y	T	R	Y	Q	T	L	E	L	E	K	E	F

Position	21	22	23	24	25	26	27	28	29	30	31	32	33	34	35	36	37	38	39	40
DNA	CAC	TTC	AAT	CGC	TAC	TTG	ACC	CGT	CGG	CGA	AGG	ATC	GAG	ATC	GCC	CAC	GCC	CTG	TGC	CTC
Protein	H	F	N	R	Y	L	T	R	R	R	R	I	E	I	A	H	A	L	C	L

Position	41	42	43	44	45	46	47	48	49	50	51	52	53	54	55	56	57	58	59	60	61
DNA	ACG	GAG	CGC	CAG	ATA	AAG	ATT	TGG	TTC	CAG	AAT	CGG	CGC	ATG	AAG	TGG	AAG	AAG	GAG	AAC	AAG
Protein	T	E	R	Q	I	K	I	W	F	Q	N	R	R	M	K	W	K	K	E	N	K

Figure 3.5
The Antennapedia *gene and the homeobox. The* Antennapedia *gene (top) consists of eight exons (blocks of messenger RNA coding sequences). The protein coding sequences are confined to the last four exons (cross-hatched), including the homeobox (black). The homeobox becomes translated into the homeodomain located near the COOH-terminus of the Antennapedia protein. The homeobox consists of 180 base pairs encoding 60 amino acids listed underneath the respective codons in the DNA. For abbreviations, see fig. 1.5.*

with 75–77 percent sequence identity between *Antennapedia, fushi tarazu,* and *Ultrabithorax.* Furthermore, the three homeoboxes share the same open reading frame for translation of the coding DNA strand into a similar protein sequence, which indicated that the homeobox encodes a protein segment that we designated the homeodomain. The sequence homology between *Antennapedia, fushi tarazu,* and *Ultrabithorax* was found independently by Matthew Scott and Amy Weiner, who had carried out a similar chromosome walk across the *Antennapedia* locus.

The finding of homologies between different members of a gene family was in itself not such an important discovery, but Bill McGinnis and I went immediately ahead and used the homeobox as a probe to clone other homeobox-containing genes from *Drosophila* and to look for the corresponding genes in other organisms. The first two genes isolated from *Drosophila* that cross-hybridized with the homeobox sequences were *Deformed (Dfd)* and *abdominal-A (abd-A),* both of which were known from homeotic mutations. Therefore, the term *homeobox* was justified.

At that time Eddy De Robertis and his group were working in our department, and we held common seminars and discussion groups to keep abreast of the latest research. Our discussions were extremely lively, stimulating, and productive. At the end of one of these research seminars, De Robertis came to my office to continue the discussion, and in a flare of boldness we decided to begin looking for homeobox genes in vertebrates. We knew that insects and vertebrates develop very differently, but because De Robertis worked on frogs, it would be easy to do at least a pilot experiment. Bill McGinnis had already done what in the jargon is called a zooblot, in which DNA from various organisms was cut with restriction enzymes, separated into fragments of various length on a gel, blotted onto a filter paper, and hybridized with a homeobox probe. These blots clearly indicated that homeoboxes might be present in DNA from various animals, including a beetle and an earthworm. A gel with frog DNA also gave some cross-hybridizing bands, but the hybridizing sequences had to be cloned and sequenced to determine that they were indeed homeoboxes. Several collaborators considered this project extremely risky and did not want to burn their fingers, but Andres Carrasco, with the substantial help of Eddy De Robertis and Bill McGinnis, cloned the first vertebrate homeobox gene from the frog *Xenopus.* This surprising finding made the headlines: it suggested that there was a common underlying principle in the genetic control of development of both vertebrates and invertebrates.

When Frank Ruddle came for a sabbatical year to my laboratory, he had planned to spend most of his time reading and writing, but when he saw what was going on around him, he couldn't resist, and he returned to the laboratory bench. Together with McGinnis, he cloned the first mouse homeobox gene and launched a new era in mammalian genetics. Not long after that the first human homeobox gene was cloned by Michael Levine after he returned from Basel to the United States (fig. 3.6).

These first experiments seemed to indicate that homeoboxes were present in segmented animals only, because the zooblots revealed no signals from the DNA of sea urchins and of the nematode worm *Ascaris,* animals that lack overt segmentation. In publishing our findings, however, I noted that at least 60 percent sequence identity is required for the detection of a positive signal, so that our failure to detect homeoboxes in sea urchins and nematodes did not mean that these animals lacked homeoboxes. In fact, with more sensitive methods, Thomas Bürglin later found a large number of homeobox genes in nematodes, and they were also isolated from sea urchins.

```
        1        10        20        30        40        50        60
Fly     RKRGRQTYTRYQTLELEKEFHFNRYLTRRRRIEIAHALCLTERQIKIWFQNRRMKWKKEN
Moth    ............................................................   100%
Bee     .....................Y......................................    98%
Plan    H..S...................K..............S...................DH    90%
Urch    ..........A..........Y......K......Q.V..S..................R    88%
Asc     S..T.TA..............Y...........S..........................    90%
Lin     .......................K....................................    98%
Amph    .......................K....................................    98%
Frog    ...........................................................H    98%
Chick   ...........................................................H    98%
Mouse   ...........................................................H    98%
Man     ...........................................................H    98%
```

Figure 3.6
Homeodomain sequences of Drosophila *Antennapedia and related homeodomain proteins from other animal species and humans. A dot indicates that the same amino acid as in* Drosophila *is found at this position.* Abbreviations: *Fly* Drosophila melanogaster *(fruit fly); Moth* Bombyx mori *(silkmoth); Bee* Apis mellifera *(honey bee); Plan* Dugesia tigrina *(planarian flatworm); Urch* Tripneustes gratilla *(sea urchin); Asc* Phallusia mammillata *(sea squirt); Lin* Lineus sanguineus *(nemertean worm); Amph* Branchiostoma floridae *(amphioxus); Frog* Xenopus laevis *(clawed frog); Chick* Gallus gallus *(chicken); Mouse* Mus musculus *(mouse); Man* Homo sapiens *(human).*

The fact that in *Drosophila* the homeobox was found primarily in genes that specify segmental identity or are otherwise involved in segmentation suggested that the homeobox might perform a similar function in vertebrates. Therefore we speculated: "If the conserved homeodomain in fruit flies, frogs, mice, and humans is involved in the control of segmental development, then it is possible that the segmentally organized animals in both the protostome and deuterostome classes had a common ancestor, and that the metameric body plan has evolved only once in metazoa." This bold speculation did not remain uncontested; Elizabeth and Rudolf Raff proposed a much more "mundane" function for the homeodomain in the journal *Nature.* The Raffs suggested that it might simply be a nuclear targeting signal, that part of a protein that allows it to enter the cell nucleus. "While we agree that protostomes and deuterostomes must share at some point a common ancestor," they wrote, "the chordates are, in fact, not segmented animals and are fundamentally distinct from arthropods [which include insects] and other metameric [=segmented] animals both in body plan and in the developmental processes by which the body plan is achieved. The evolution of these developmental programs offers little evidence of meaningful homology." Since then, however, we have accumulated more and more evidence that the same homeobox genes are used in both vertebrates and invertebrates to specify the body plan and that the mechanisms of the genetic control of development are much more universal than anticipated. As Stephen Jay Gould has pointed out, our speculative hypothesis was proposed as early as 1830 by Etienne Geoffroy Saint-Hilaire and discussed in a public debate at the Académie des Sciences in Paris. At that time the arguments of Geoffroy Saint-Hilaire were defeated by the more eloquent Georges Cuvier. Now, a hundred and fifty years later the pendulum has swung in the opposite direction. I never felt that scientific questions could be solved in public debates.

The first hint about the function of the homeodomain came from a computer search through protein data bases carried out by John Shepherd in our laboratory. Surprisingly, Shepherd found a small but significant degree of sequence identity between the mating-type proteins of yeast and the homeodomain (fig. 3.7). The sequence identity ranged only between 22 percent and 28 percent and could not have been detected by hybridization, but it appeared statistically significant. This was a critical finding, because the yeast geneticists had shown previously that the MAT genes, which control the mating type, are regulatory genes and that MATα2 encodes a repressor pro-

Sequence Homology between Drosophila Antennapedia and MAT homeodomains in Yeasts

RIEIAHALCLTERQIKIWFQNRRMKWKKEN
KEEVAKKCGITPLQVRVWVCNMRIKLKYIL 14 / 60
KEEVAKKCGITPLQVRVWFINKRMRSK* 14 / 57
LENLMKNTSLSRIQIKNWVSNRRRKEKTIT 17 / 60
FYDLSAATGLTRTQLRNWFSNRRR* 13 / 54

Figure 3.7
Sequence homology between Drosophila Antennapedia *and the yeast mating type genes. The amino acid sequences between the homeodomains of* Antennapedia *and the mat proteins of yeast are compared: mat a1, mat a1 spliced variant, and mat α2 are from baker's yeast* (Saccharomyces cerevisiae)*; mat P1 is from fission yeast* (Schizosaccharomyces pombe). *The ratios (e.g., 14/60) indicate the number of amino acids identical to ones in* Antennapedia. *Identical (conserved) amino acids are boxed.*

tein that represses the genes specific for the a mating type, resulting in the formation of a cell of the α mating type.* Thus, MATα2 is involved in determining cell type and encodes a gene regulatory protein. The partial homology between the MAT and the homeodomain proteins suggested a similar function for homeodomain proteins, as Alan Garen and I had expected all along. As I shall demonstrate later, both of these hypotheses were correct: yeast does have homeodomains, and homeodomain proteins are gene regulatory proteins that regulate batteries of target genes.

Our finding of putative homeodomains in yeast was taken by the Raffs as a crucial counterargument against our hypothesis that the homeobox genes specify the segmental body plan, since "certainly there is little argument for *Saccharomyces* as a segmented organism" (yeast is a single-celled organism and reproduces by budding). Other colleagues were highly skeptical, too; at

*Mating types in yeast are the equivalent of the sexes in higher organisms. Only cells of different mating type (a or α) mate to form a heterzygous a/α diploid cell.

an international meeting in which I presented these results, Eric Davidson sat in the front row and conspicuously shook his head during most of my talk, showing his disapproval of my wild hypothesis. A few years later, however, he became a convert and was happily cloning homeobox genes from his beloved sea urchins.

How did the press respond to these discoveries? The editor of *Nature* who handled our first paper liked our findings, but he disliked the term *homeobox* and replaced it with *homoeotic sequence* just before the manuscript went to press. But he forgot to remove it from a later part of the article, so the term *homoeobox* (still in the old spelling) did appear in our first publication. (This incident illustrates the power journalists can exert on science though they are not qualified to make such judgments.) The *New York Times* published an article entitled "Human and Insects Appear to Share Fragment of Gene" in which Gary Struhl, professor of biochemistry at Harvard, described the discovery of the homeobox and added that "it may prove to be a major breakthrough in understanding vertebrate development." Struhl's name was mentioned several times, even though his involvement in the discovery was limited to having kindly provided us with some mutants, and then two other American scientists were cited extensively, although they, too, did not contribute to the discovery of the homeobox in vertebrates. Finally, in the last five lines of the article the authors of the Basel group were mentioned. I was used to this kind of nationalism in sports. Science should be free of such chauvinism, however, and I should like to note here that two Swiss, two Americans, and one Japanese were coauthors of our first paper describing the homeobox. The Swiss newspaper *Die Weltwoche* countered the *New York Times* by claiming that "the key to life lies in Basel," and one English writer even compared the homeobox discovery to the Rosetta stone, yet we still have a long way to go before we will have deciphered the developmental program in DNA.

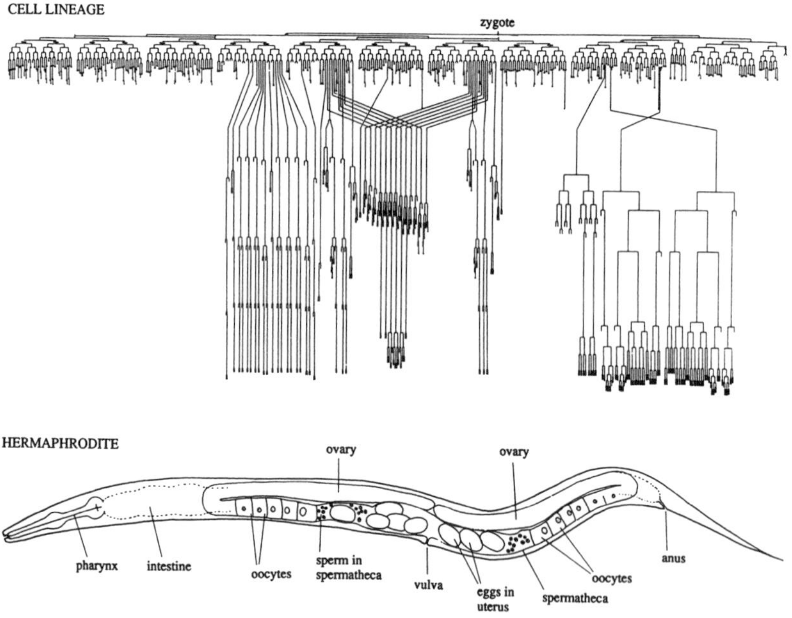

Figure 4.1

Cell lineage and anatomy of the hermaphrodite of the nematode worm Caenorhabditis elegans. *Each vertical line corresponds to a cell. Anterior is to the left, posterior to the right. Most cells are born near their final position, but some emigrate anteriorly or posteriorly over considerable distances (crossing lines). From J. E. Sulston et al., The embryonic cell lineage of the nematode* Caenorhabditis elegans, Developmental Biology *100 (1980): 64–119, and W. B. Wood, "Sexual Dimorphism and Sex Determination," in Wood et al.,* The Nematode "Caenorhabditis elegans" *(New York: Cold Spring Harbor Laboratory, 1988).*

~

4
How to Generate an Embryo?

To watch fertilization with your own eyes directly under the microscope is one of the most fascinating experiences anyone interested in life sciences can have. I provide this opportunity to my students by teaching a marine embryology course in Banyuls on the Mediterranean coast in southern France. Each student is supplied with a dish of eggs and a dish with sperm from a pair of sea urchins. The students add a drop of diluted sperm solution to a suspension of eggs in sea water and observe under the microscope how the sperm enters the egg and how an embryo is generated. Nature is extremely generous in producing gametes: eggs are produced by the millions and sperm by the billions in a vast excess to ensure reproduction, even though only a tiny fraction of the gametes is successful. The examination of sea urchin fertilization under the microscope dates back to Oskar Hertwig in 1875; sea urchins have transparent eggs, and artificial fertilization is easily accomplished. Nowadays, the nuclei of egg and sperm and the chromosomes can be visualized in other living cells by fluorescent dyes that do not interfere with development. It is amazing to watch as thousands of sperms swarm around an egg covered by a coating of jelly. Only one sperm is the lucky winner, penetrating the jelly coating fastest and reaching the egg first. The membranes of the two cells immediately fuse, and the sperm nucleus moves toward the center of the egg, where it meets the egg nucleus and the two nuclei fuse. Because the egg and sperm nuclei each contain one set of chromosomes, the fused nucleus contains two sets of chromosomes—that is, two haploid cells unite to form a normal diploid cell containing a maternal and a paternal set of chromosomes. The successful sperm penetration can be detected by the formation of a "fertilization membrane" that lifts from the surface of the egg and forms a barrier to prevent further sperm entries. Within about an hour the egg cleaves into two cells and embryogenesis begins.

There are two extreme ways to generate an embryo; Sidney Brenner has designated them the European way and the American way. The European way is for the cells to do their own thing and not to talk to their neighbors very much. Ancestry is what counts, and once a cell is born in a certain place it will stay there and develop according to rigid rules; it does not care about the neighborhood, and even its death is programmed. If it dies in an accident, it cannot be replaced. The American way is quite the opposite. Ancestry does not count, and in many cases a cell may not even know its ancestors or where it came from. What counts are the interactions with its neighbors. It frequently exchanges information with its fellow cells and often has to move to accomplish its goals and find its proper place. It is quite flexible and competes with other cells for a given function. If it dies in an accident, it can readily be replaced.

"The Worm": The European Mode of Development

In many respects the paradigm for a European mode of development is *Caenorhabditis elegans*, a small nematode worm, the favorite organism of Sidney Brenner (though recent work has revealed some American traits as well). "The worm," as the insiders call it, was deliberately selected around 1963, when Brenner and Francis Crick felt strongly that most of the classical problems of molecular biology had been solved and that the future lay in tackling more complex biological questions: "We must move on to other problems of biology which are new, mysterious and exciting." Brenner chose the development of the worm and Crick began to tackle the human brain. Why the worm? Brenner wanted to tame a small, simple metazoan organism to study development directly by subjecting it to the analytical methods of microbial genetics. He chose *C. elegans*, a small—about one millimeter long—free-living soil nematode that feeds primarily on bacteria. It reproduces with a life cycle of about three days and produces several hundred progeny. It is ideally suited for genetics because it can be grown on agar plates carrying a lawn of *Escherichia coli* bacteria. It comes in two "sexes," hermaphrodites, which produce both eggs and sperm and can reproduce by self-fertilization, and males, which arise spontaneously at low frequency. Males can also fertilize hermaphrodites, allowing the experimenter to set up genetic crosses. The haploid genome size is 8×10^7 base pairs, about half of that of the fruit fly, and the project to decipher the worm's entire genome is well under way. *C. elegans* is also anatomically relatively simple, with a fixed number of cells that can be counted by direct observation under the microscope. There are only 959 somatic cells in the adult hermaphrodite and 1,031 in the adult male plus a variable number of germ cells.

The cell divisions of *C. elegans* can also be observed microscopically in the living animal, and its complete cell lineage, the route by which all the cells of the adult are derived from the zygote (the fertilized egg) by cell division, has been determined (fig. 4.1). The cell lineage of *C. elegans* is essentially invariant from animal to animal, and the organism is generated with a high degree of precision.

In principle, three basic modes of cell division can be distinguished: (1) proliferative division, which is symmetrical and leads from a mother cell to two daughter cells of the same type (A → A+A); (2) stem cell division, in which a stem cell A divides repeatedly asymmetrically into a cell A (which remains a stem cell) and a cell B that differentiates (A → A+B); and (3) diversifying division, in which a cell A leads to two cells differing from both their mother and their sister cell (A → B+C).

The first four divisions in the *C. elegans* embryo are stem cell divisions in which the P-cell remains a stem cell and splits off the four founder cells AB, EMS, C, and D (fig. 4.2). The P-cell is the primordial germ cell from which all germ cells (eggs and sperm) are later derived. The EMS cell undergoes a further diversifying division and gives rise to the founder cells MS and E. The founder cell E undergoes the most simple pattern of cell division and generates a single clone of twenty gut cells, which is known as monoclonal development. The lineage of founder cell C is more complicated, because in addition to proliferative divisions there are also diversifying divisions, which give rise to cell progeny with different developmental fates—muscle cells, skin cells, nerve cells, and a cell that undergoes cell death. No fewer than 113 out of 1,031 cells undergo programmed cell death, a surprisingly large num-

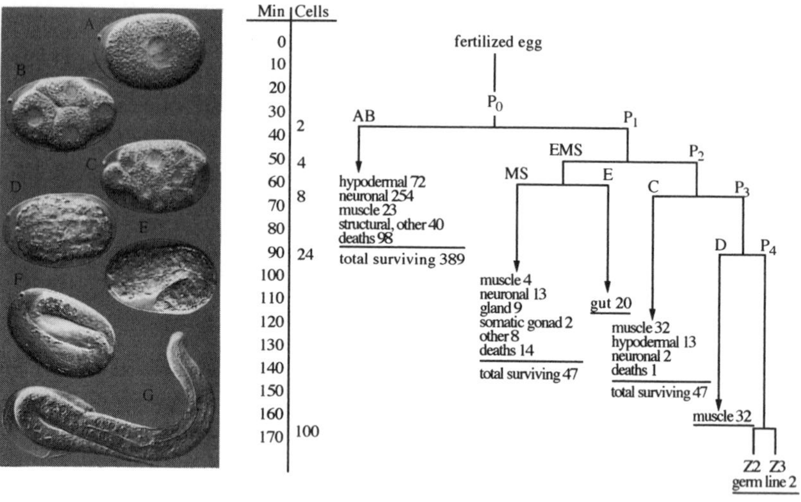

Figure 4.2
Embryogenesis and early cell lineage of C. elegans. Left: *Microphotograph of embryonic stages; courtesy of R. Schnabel.* Right: *Cell lineage of the germline founder cells (P_0–P_4), the somatic founder cells AB, MS, E, C, D, and the number and fate of their progeny; after W. B. Wood, "Embryology," in Wood et al.,* The Nematode "Caenorhabditis elegans" *(Cold Spring Harbor, N.Y.: Cold Spring Harbor Laboratory, 1988), 220.*

ber. Cell death, like muscle cell differentiation, is a developmental program as much as natural death is part of the life cycle. In most cases the cell destined to undergo cell death commits suicide and dies without external influences. Mutations that block this mechanism, which is also called apoptosis, are detrimental and sometimes lethal for the organism. In some cases a cell may also be killed by a neighboring cell and engulfed by other cells, but this is rare.

In *C. elegans* cell lineage generally decides a cell's fate; it is the ancestry that counts, as expected for the European way of development. The fate of a nucleus depends on the region of the egg cytoplasm that it occupies and on its neighbors. If an individual cell is killed with a laser microbeam, it is not replaced, and the respective cell and its progeny are missing from the adult worm. Mutations affecting cell lineage do also affect the fate of the respective cells, indicating a causal relation between cell lineage and cell fate. This programmed mode of development with a limited capacity of repair is successful only when combined with a large number of progeny; it is somewhat reminiscent of the mass-produced disposable items in today's consumer culture. Nevertheless, even in the worm, cellular interactions play an important role in development, because the fixed cell lineage depends on constant cell contacts and thus on cellular interactions.

How do cells acquire different developmental fates? How do cells differentiate, as the developmental biologists say? Since the early twentieth century two mechanisms of cell differentiation have been primarily discussed; one is based on factors intrinsic to the cell leading to autonomous differentiation, and the other involves interactions among cells, one cell inducing another cell to acquire a certain developmental fate. The earliest steps in cell differentiation, when a developmental pathway is initiated, are called determination. In the case of the intrinsic, cell-autonomous mechanism, cytoplasmic substances (determinants) are thought to be differentially distributed among the daughter cells, and the particular cell that receives the determinants becomes programmed for a certain developmental fate.

Only a few clear-cut examples of such determinants exist. Among the better characterized ones are the germ cell determinants of the worm and of *Drosophila* (as described later). In *C. elegans,* the *cib-1* mutation (*cib* stands for *c*hanged *i*dentity of early *b*lastomere) blocks the determination of the P1, P2, and P3 cells, which are supposed to undergo three asymmetric stem cell divisions (see fig. 4.2). The egg—that is, the Po cell—contains germline-specific particles (P-granules) in its cytoplasm, at the posterior pole of the egg. These particles are differentially distributed among the two daughter cells and invariably accumulate in the P-cell, which eventually gives rise to all the germ cells. In the *cib-1* mutant embryo no typical P1 cell is formed, and it loses its ability to divide asymmetrically like a stem cell; instead, it divides such that the P-granules are distributed equally among the daughter cells. The mutant P1 cell omits the next cell division and subsequently assumes the same fate as its normal daughter cell, so that it also becomes an EMS founder cell. The product of the normal *cib-1* gene is therefore a good candidate for a determinant, a germ cell determinant in this case, or it might control the localization of the determinants.

Even though the worm has a fixed pattern of cell division, cellular interactions play a major role in cell differentiation. Because each cell division is precisely programmed, the positions of the cells relative to one another are also initially fixed, and inductive interactions are not immediately obvious. However, such interactions can be demonstrated, for example, by destroying a given cell with a laser microbeam and examining the fate of the surrounding cells. In this way, it has been demonstrated that only the antero-posterior axis is predetermined in the nematode egg by cytoplasmic determinants, deposited in defined regions of the egg cytoplasm. In contrast, the left-right asymmetries are determined by inductive interactions among cells. The best-studied case of *inductive interactions* concerns the formation of the vulva in the hermaphrodite worm. The vulva (see fig. 4.1), required for egg laying, develops from six vulval precursor cells that are located within the ventral epidermis. Although in normal development each vulval precursor cell assumes a particular developmental fate, a variety of experiments indicate that each precursor is equivalent in its ability to assume either of three different fates and thus that the six precursors constitute an equivalence group. Each fate consists of a distinct cell lineage that produces a particular set of cell types (fig. 4.3). Fates 1° and 2° produce vulval tissue contributing to different regions of the vulva, whereas fate 3° gives rise to nonvulval epidermis. The fate of the precursors is determined by three intercellular signals. The adjacent anchor cell of the gonad produces a signal that induces the closest precursor cells to assume fates 1° and 2°. A lateral signal among the induced precursor cells regulates the pattern of fates 1° and 2°, and an inhibitory sig-

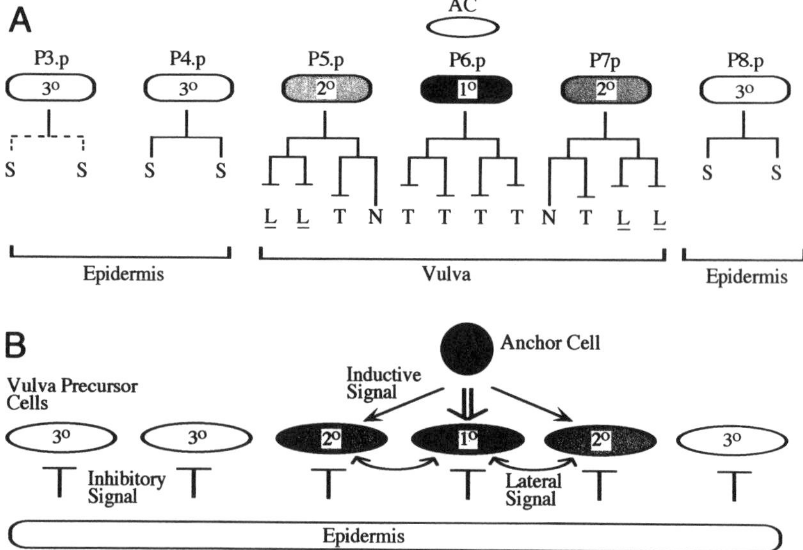

Figure 4.3
Vulva induction in C. elegans: *(A) cell lineage and primary, secondary, and tertiary cell fates; (B) inductive signals (arrows) and inhibitory signals (T-bars) between the different cell types. From R. J. Hill and P. W. Sternberg, Polarity and adhesion in development, in P. Ingham, A. Brown, and A. Martinez Arias, eds.,* Development Supplement 1993 *(Cambridge: Company of Biologists, 1993), 9–18.*

nal emanating from the epidermis prevents the precursors from assuming a vulval fate in the absence of the inductive signal. In normal development the combined action of the three signals determines the precursor cells to differentiate according to the pattern 3° 3° 2° 1° 2° 3°.

The classical embryological approach to study cell determination consists of removing a cell or a group of cells by cell ablation and to study the effects upon the remaining cells of the embryo (table 4.1). If the anchor cell is removed by laser ablation (destruction), the vulval precursor cells that would normally assume the 1° and 2° fates instead assume fate 3° and no vulva is formed, indicating that the anchor cell is necessary for vulva induction. If cell P6.p, which normally would assume the 1° fate, is ablated, its neighboring cell P5.p will assume the 1° fate because it belongs to the same equivalence group. The adjacent P4.p cell is also respecified and assumes the 2° fate, resulting in an essentially normal 3° 2° 1° 2° 3° pattern of vulva differentiation except that one cell is missing.

A second classical embryological approach consists of transplanting a given cell to the same site of another embryo (homotopic transplantation) or to a different site (heterotopic transplantation) in order to study the determinative interactions with neighboring cells. Single cell transplantations are harder to carry out in the worm than in some other organisms, but *C. elegans* is highly suitable for genetics. In *dig-1* mutants the anchor cell can be shifted anteriorly, so that it lies close to the P5.p rather than P6.p cell. This shift represents an anterior "transplantation" of the anchor cell and results in the induction of the 1° fate in the P5.p cell rather than in P6.p, and a pattern 3° 2° 1° 2° 3° 3° is formed that is shifted anteriorly by one cell. By this mechanism of induction it is ensured that the 1° cell is formed right next to the anchor cell and not somewhere else. Cell autonomy versus cellular interactions can also be studied by generating genetic mosaics of mutant and wild-type cells, which in addition allow the experimenter to study clones of genetically marked cells.

The enormous power of the genetic approach is to identify the genes that regulate development by mutation, to isolate the key genes by recombinant DNA technology, and to study the structure and function of the respective gene products, which eventually leads to an understanding of the molecular basis of development. Genetic techniques have allowed the identification of the key players of vulva induction—that is, the signal substance produced by the anchor cell and the respective receptor on the precursor cells (see table 4.1). The *lin-3* gene encodes the inductive signal. Mutations in *lin-3* that reduce its function have a vulvaless phenotype in which all three of the vul-

Table 4.1
Vulval fate patterns

Constitution	Fate of vulva precursor cells					
	P3.p	P4.p	P5.p	P6.p	P7.p	P8.p
Normal (wild type)	3°	3°	2°	1°	2°	3°
Ablation of P6.p	3°	**2°**	**1°**	X	2°	3°
Ablation of anchor cell	3°	3°	**3°**	**3°**	**3°**	3°
Vulvaless mutant	3°	3°	**3°**	**3°**	**3°**	3°
Multivulva mutant	**1°**	**2°**	**2°**	1°	2°	1°
dig-1 mutant (anterior shift of anchor cell)	3°	**2°**	**1°**	**2°**	**3°**	3°

AFTER R. J. HILL AND P. W. STERNBERG, "Cell fate patterning during *C. elegans* vulval development," *Development* (1993) suppl.: 9–18.

val precursor cells that usually assume vulval fates 1° and 2° instead assume the epidermal fate (3°). In contrast, transgenic worms that carry many additional copies of the normal *lin-3* gene (gain-of-function mutants) show a multivulva phenotype in which all six vulva precursor cells may assume vulval fates. This indicates that the normal (wild type) *lin-3*$^{+}$ gene is both necessary and sufficient to induce the vulval fate. As expected, *lin-3* is expressed specifically in the anchor cell from which the signal emanates. *lin-3* encodes a member of the epidermal growth factor (EGF) family, which includes EGF, a protein that stimulates the growth of epidermal cells in mammals, and other ligands, proteins that bind to EGF receptors (fig. 4.4). Proteins of the EGF family are made as membrane-spanning proteins that contain at least one extracellular EGF domain. This sequence motif consists of approximately fifty amino acids with six cysteine residues that form three intramolecular sulfur bridges, giving it a rigid structure. In the case of the lin-3 protein, the EGF domain is probably cleaved from the rest of the protein to produce a secreted factor, like a hormone, that can diffuse to its neighboring cells and activate its receptor on their surface. Therefore, the induction of the vulval fate can occur without direct contact between the anchor cell and the induced vulval precursor cells. In other cases such a direct contact is required.

The gene that encodes the receptor has also been identified by mutations that act downstream of *lin-3* and affect the *let-23* gene, a homologue of the EGF receptor in mammals. LET-23 is a transmembrane protein with an extracellular domain and a tyrosine kinase catalytic domain on the cytoplasmic side. Upon binding of lin-3 (the ligand) to the extracellular domain of *let-23* (the receptor), the catalytic domain becomes activated by phosphorylation of specific amino acid residues. The activated receptor catalyzes the phosphorylation of tyrosine residues in other proteins and initiates an intracellular signal transduction cascade, which transmits the signal from the receptor on the surface of the cell to the nucleus. Upon receiving the signal, the nucleus responds by activating or repressing specific target genes. In this way, the inductive signal is transmitted from the anchor cell to the vulval precursors.

Sea Squirts: The Yellow Crescent and Half Larvae

Other animal phyla have different modes of development, but as we shall see later, there are common underlying mechanisms and recurring themes among the various developmental programs leading from the egg to the adult organism. Like nematodes, Ascidians (sea squirts) have a fixed cell lineage. The normal development of the sea squirt *Phallusia mammilata* from the fertilized egg to the tadpolelike larva is shown in figure 4.5. As early as 1905, Edwin Conklin determined the cell lineage of the *Styela* embryo,

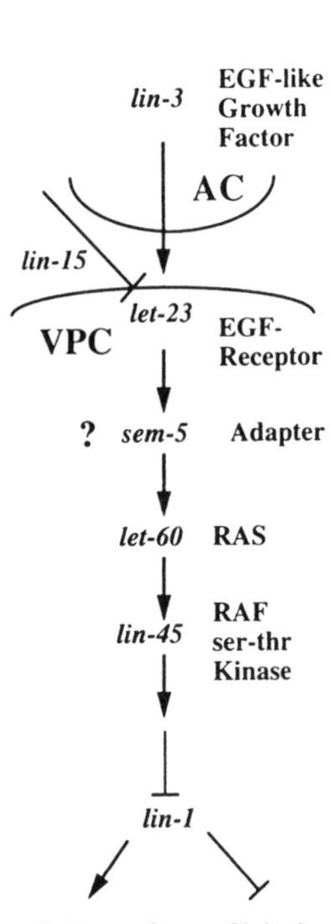

Figure 4.4
Signal transduction pathway in vulva induction. Abbreviations: AC *anchor cell;* VPC *vulva precursor cell;* lin-3, lin-15, let-23, sem-5, let-60, lin-45, lin-1 *interacting genes or their protein products;* EGF *Epidermal Growth Factor;* RAS, RAF *mammalian homologues; arrows, positive regulation (activation); T-bars, negative regulation (repression). From R. J. Hill and P. W. Sterriberg, Polarity and adhesion in development, in P. Ingham, A. Brown, and A. Martinez Arias, eds.,* Development Supplement *1993 (Cambridge: Company of Biologists, 1993), 9–18.*

another sea squirt, by direct microscopic observation and wrote the classical monograph *The Organization and Cell-Lineage of the Ascidian Egg.* The results of this beautiful study have been confirmed for *Halocynthia roretzi,* a Japanese sea squirt, by Hiroki Nishida and Noriyuki Satoh using modern techniques. Single blastomeres can be microinjected with a stable innocuous marker enzyme such as horseradish peroxidase, which is passed on from the injected cell to its progeny. Because the enzyme cannot penetrate the neigh-

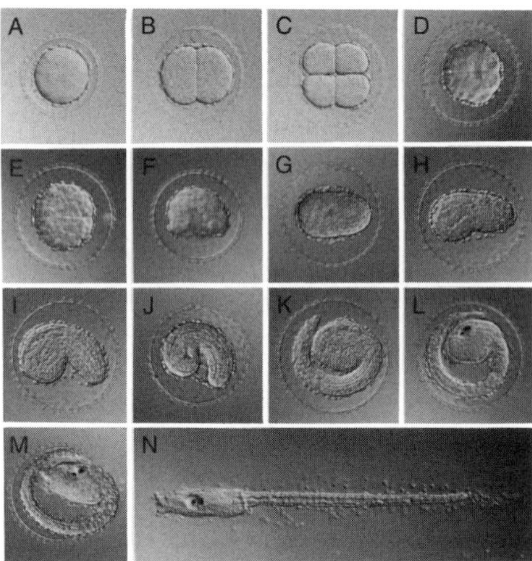

Figure 4.5
Embryonic development of the sea squirt Phallusia mammillata: *(A) egg; (B) two-cell stage; (C) four-cell stage; (D) early blastula; (E) late blastula; (F) gastrula; (G) early neurula; (H) late neurula; (I) early tailbud stage; (J) middle tailbud stage; (K) late tailbud stage; (L) early larva; (M) hatching larva; (N) swimming larva. Courtesy of S. Glardon.*

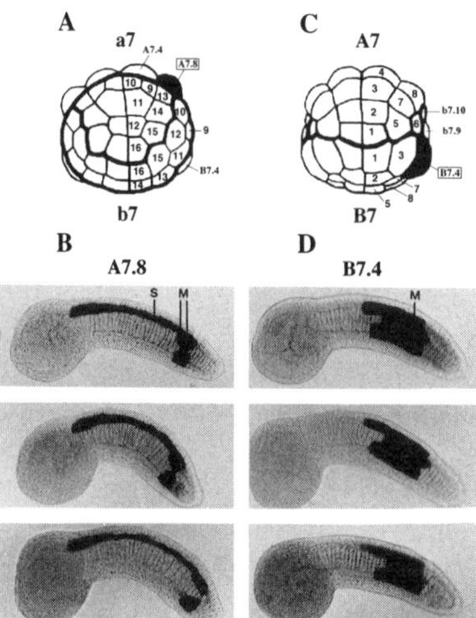

Figure 4.6
Fixed cell lineage in the sea squirt Halocynthia roretzi: *(A) injection of cell A7.8 at the blastula stage with a tracer; (B) identical clones comprising spinal chord (S) and muscle cells (M) at the tailbud stage of three injected larvae; (C) tracer injection into cell B7.4; (D) identical clones of muscle cells (M) in the tail of three injected larvae. From H. Nishida, Cell lineage analysis of ascidian embryos by intracellular injection of a tracer enzyme,* Developmental Biology *121 (1987): 526–41.*

boring cells, the descendants of the injected cells can be labeled with a color reaction revealing peroxidase activity. In this way a clone of marked cells can be visualized in the embryo at a later stage and the cell lineage can be precisely determined. If the same cell is injected in different embryos, the resulting clones invariably form the same pattern, reflecting the invariant cell lineage (fig. 4.6). The cell lineage found in *Halocynthia* is essentially identical to the one of *Styela* as described by Conklin some eighty years earlier.

The unfertilized *Styela* egg studied by Conklin is radially symmetrical with respect to the animal-vegetal axis. The egg nucleus is located near the animal pole, whereas the opposite pole, called vegetal, is rich in yolk. Upon entry of the sperm, a spectacular rearrangement of the egg cytoplasm occurs that determines the dorsoventral and anteroposterior axes of the future embryo. Because the different regions of the egg cytoplasm are differentially colored, Conklin was able to assess the relation between the cytoplasmic movements and the embryonic axes. The cortical cytoplasm around the periphery of the egg contains bright yellow pigment granules whose movements can be followed easily in the living egg (pl. 2). During the first phase of cytoplasmic movements, a wave of contraction beginning at the animal pole sweeps across the egg and leads to the accumulation of the yellow cortical cytoplasm at the vegetal pole, where it forms a yellow cap. The sperm tends to enter into the animal hemisphere and is also moved by the contraction wave toward the vegetal pole. During the second phase of cytoplasmic movements, the yellow cytoplasm, together with the sperm nucleus, shifts from the vegetal pole toward the equator of the egg and occupies a subequatorial region, forming what is called the yellow crescent (pl. 2). The yellow crescent marks the future posterior region of the embryo, and the yellow cytoplasm becomes segregated into those cells that later will form the muscle cells of the tail in the developing tadpole. The nuclei of sperm and egg fuse in the center of the egg. The first cleavage furrow exactly bisects the yellow crescent and divides the egg precisely into two symmetrical cleavage cells (blastomeres), one giving rise to the left side of the tadpole, the other to the right side.

In 1887, Laurent Chabry published the first report on microsurgical experiments with ascidian embryos. He undertook the first half-blastomere isolation experiments in which he destroyed one of the first two blastomeres, either the left or the right cell, and obtained "half-larvae." These "demi-individu droit" or "demi-individu gauche," as he called them, developed quite far, forming a tail, a pigment spot, and a palp, or adhesive organ. That he observed one pigment spot and one palp only clearly indicates that the experimental tadpole was a "half-larva," because the normal larva has two pigment spots, the ocellus (eye) and the otolith (gravity sensory organ) and three palps, which adhere to the substrate during metamorphosis from the tadpole to the adult sea squirt.

Chabry's experiment was repeated in a more quantitative way by Arthur Cohen and Norman Berrill (1936), who found that half-blastomeres give rise to tadpoles with approximately half the number of muscle cells, half the number of notochord cells, and half the number of palps; the number of pigment spots, however, is not always one but varies between zero and one or two. The ocellus and the otolith are associated with pigment cells that derive from a pair of bilaterally symmetrical blastomeres at the sixty-four-cell stage and are equivalent. Both can still give rise to pigment cells of either the ocellus or the otolith. The two precursor cells then migrate anteriorly to a median position, and the more anterior cell differentiates into the pigment cell of the otolith, whereas the more posterior cell becomes part of the ocellus. Presumably, the migrating cells receive positional information from the surrounding cells and differentiate accordingly. The concept of positional information in connection with homeotic genes is discussed more extensively later. In the present context it is important to note that the isolated half-blastomeres differentiate largely as if they were still in the intact embryo, but they do have a limited capacity for regulation and can form, for example, an additional sensory organ, such as an ocellus or an otolith.

From the cell lineage diagram we can deduce when a cell becomes committed to a certain developmental pathway, when it becomes determined, as the embryologists say. At this time all the progeny of the cell enter the same pathway and have the same developmental fate. The time of cell fate determination differs in different lineages, reflecting a stepwise determination (fig. 4.7). An alternative method to cell lineage analysis for finding out whether a cell is determined is to transplant the cell to a different place in the embryo and to find out whether it develops autonomously or whether it responds to external cues and develops according to its new position. This experiment is harder to do in ascidians but is discussed in more detail later in connection with germ cell determination in *Drosophila.*

The Sea Urchin Embryo: Half-Blastomeres and Morphogenetic Gradients

When Hans Driesch in 1891 isolated half-blastomeres from sea urchin embryos, he obtained entirely different results from those of ascidians. If the two blastomeres of an embryo at the two-cell stage are separated, they both

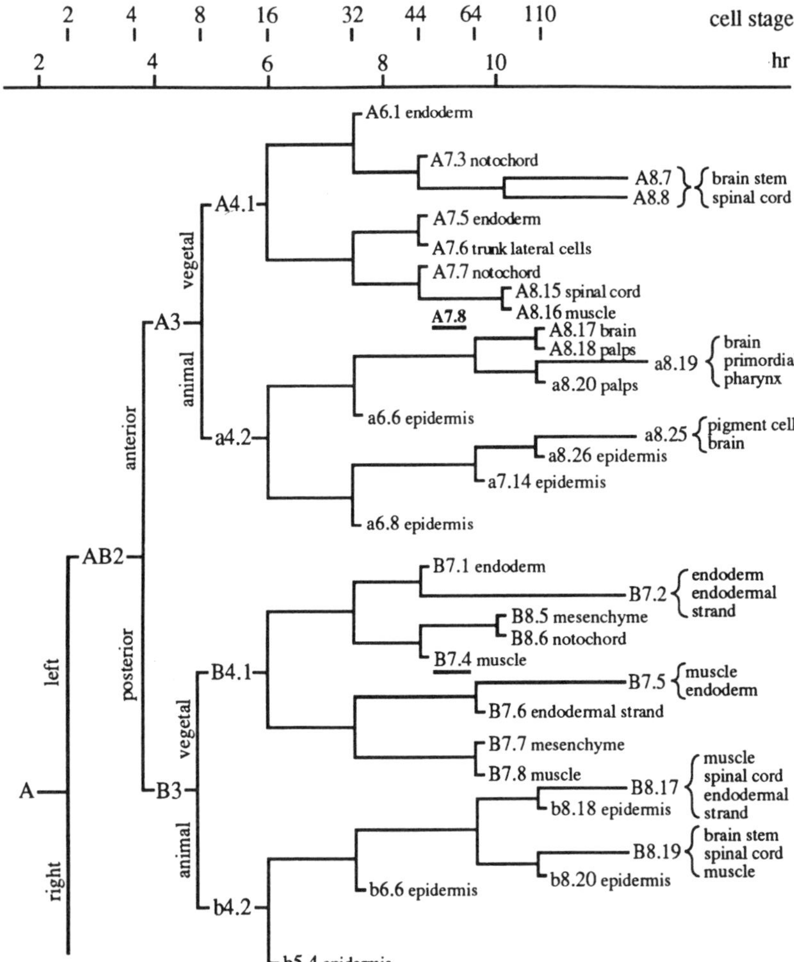

Figure 4.7

Cell fate determination in the sea squirt Halocynthia roretzi. *Only the left half of the cell pedigree is shown. The cell fate becomes determined (fixed) at various stages; the endodermal fate, e.g., is fixed in cell A6.1 at the 32-cell stage, whereas the notochordal fate determination of cell A7.3 occurs at the 44-cell stage only. From H. Nishida, Cell lineage analysis of ascidian embryos by intracellular injection of a tracer enzyme,* Developmental Biology *121 (1987): 526–41.*

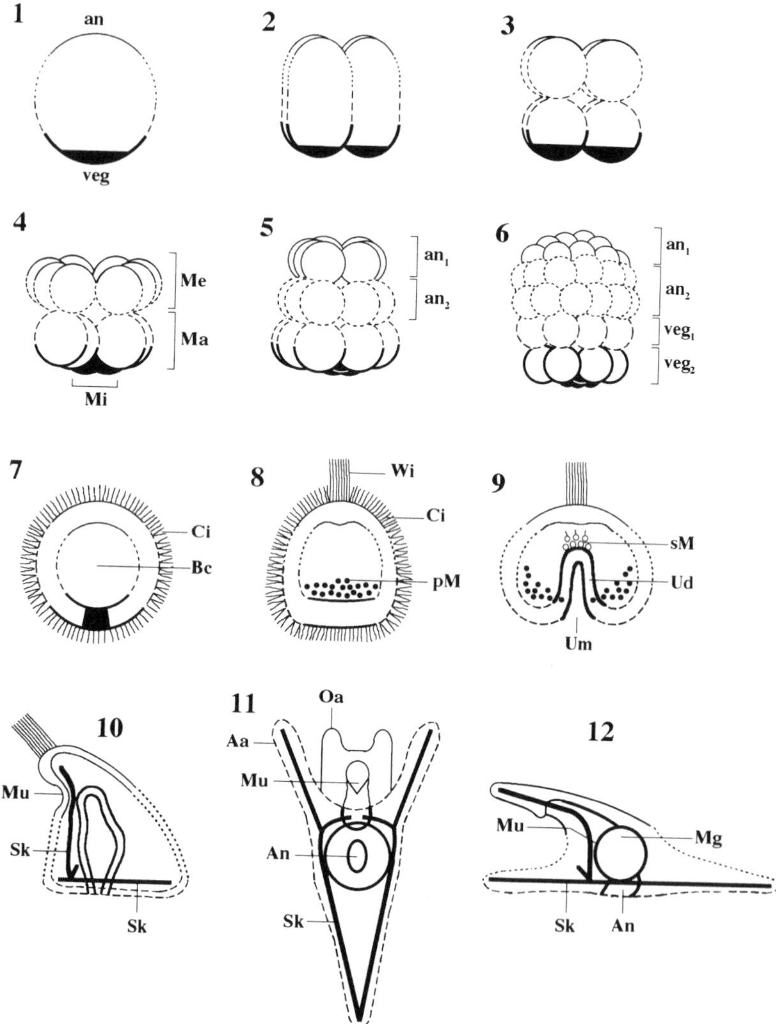

Figure 4.8
Embryonic development of the sea urchin Paracentrotus lividus: *(1) egg,* an *animal pole,* veg *vegetal pole; (2) four-cell stage; (3) eight-cell stage; (4) 16-cell stage,* Me *mesomeres,* Ma *macromeres,* Mi *micromeres; (5) 32-cell stage,* an_1, an_2 *animal tiers; (6) 64-cell stage,* veg_1, veg_2 *vegetal tiers; (7) ciliated blastula,* Ci *cilia,* Bc *blastocoel; (8) tufted blastula,* Wi *ciliary tuft,* pM *primary mesenchyme cells; (9) gastrula,* sM *secondary mesenchyme,* Ud *archenteron,* Um *blastopore; (10) prism stage,* Mu *mouth,* Sk *skeletal rod; (11) and (12) pluteus larva,* Oa *oral arms,* Aa *aboral arms,* Mu *mouth,* An *anus,* Sk *skeletal rod,* Mg *stomach. From S. Hörstadius,* Experimental Embryology of Echinoderms *(Oxford: Clarendon Press, 1973).*

develop into normal larvae, albeit of smaller size, and form identical twins. In contrast to ascidians, the left-right axis of the sea urchin egg is not fixed before the first cleavage division, and both cells remain totipotent and can give rise to complete larvae. Driesch had discovered embryonic regulation. He immediately realized the important implication of his results: if each half-egg forms a whole rather than a half-embryo, then an interaction must occur between the two cells to restrict their potential in normal development. The normal development from the fertilized sea urchin egg to the pluteus larva is schematically represented in figure 4.8. Upon fertilization, the egg undergoes two meridional divisions and then an equatorial division, which leads to eight cells of approximately equal size. During the fourth cleavage, however, the cytoplasm is distributed unequally among the cleavage cells; the four cells at the vegetal pole divide into four micromeres that are much smaller than their sister cells, the four macromeres, whereas the four cells at the animal pole divide equally to give rise to eight mesomeres. After the sixty-four-cell stage, with its two animal tiers and two vegetal tiers, the embryo forms a hollow ball of ciliated cells, called the blastula, which hatches from the egg jelly coat and swims by beating its cilia. However, the blastula is not spherically symmetrical; it retains its polarity and forms a tuft of cilia at the animal pole. At the vegetal pole, the micromeres enter the blastula and form the primary mesenchyme which subsequently forms the larval skeleton. This is the beginning of a process called gastrulation, which leads to the formation of the three germ layers: the outer ectodermal layer, the inner endoderm, and the mesoderm in between. During gastrulation, the vegetal tier 2 (veg2) invaginates around the blastopore (bp) and forms the gut. The gastrula bends over to one side and forms the bilaterally symmetrical pluteus larva, which later metamorphoses into a radially symmetrical sea urchin.

The egg is polarized before fertilization; the nucleus is located close to the animal pole (an), and the cytoplasm contains a gradient of morphogenetic substances along the animal-vegetal axis. Sven Hörstadius demonstrated this gradient in a series of classical experiments in which he cut an unfertilized sea urchin egg in half with a fine glass needle and fertilized the two fragments separately. If the egg is cut along a meridian, both halves, if supplied with a nucleus, can give rise to normal larvae, as expected from Driesch's isolation experiment with the half-blastomeres. In contrast, if the egg is cut along the equator, the two halves have different fates; the animal half forms a blastula with an enlarged tuft of cilia (see fig. 4.8) that covers almost the entire sur-

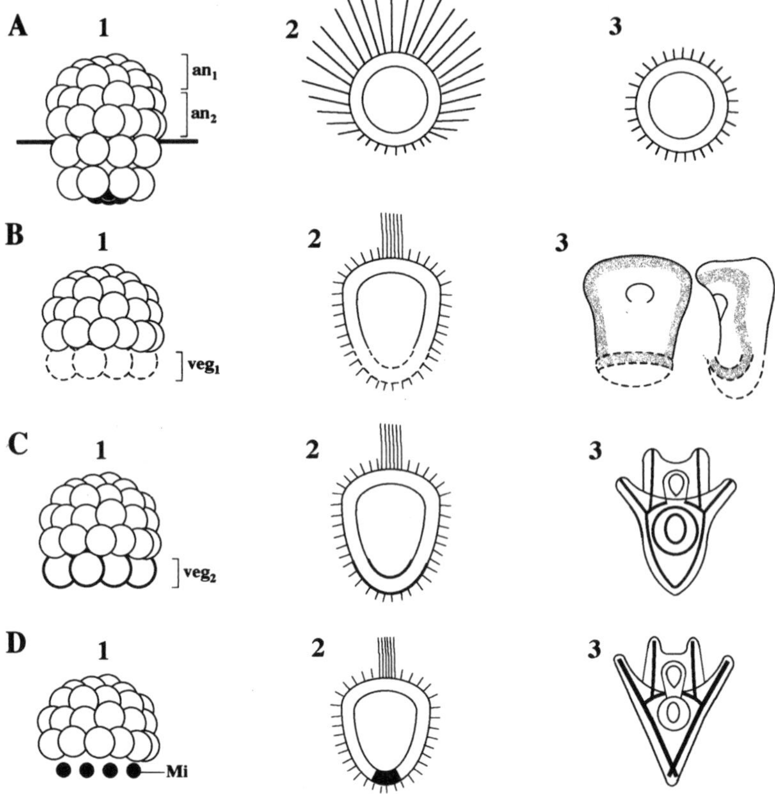

Figure 4.9

Morphogenetic gradient in the sea urchin embryo. Development of: (A) isolated an_1 + an_2 tiers; (B) an_1 + an_2 + veg_1 ; (C) an_1 + an_2 + veg_2; and (D) an_1 + an_2 + micromeres (Mi). Note the progressive normalization of development from an animalized blastula in (A) to a fairly normal pluteus larva in (D). The skeletal rods are derived from veg_2 cells in (C) but are formed by micromeres in (D). From S. Hörstadius, Experimental Embryology of Echinoderms *(Oxford: Clarendon Press, 1973).*

face. Such embryos develop neither a skeleton nor gut structures, with the exception of part of the mouth lobe. The vegetal halves are less abnormal; they develop both skeleton and gut, but they fail to form a tuft and have a reduced mouth lobe. In the course of cleavage, different regions of the egg cytoplasm are segregated into different cells. Hörstadius was able to demonstrate the existence of a morphogenetic gradient by isolating and recombining the various tiers of the sixty-four-cell stage (fig. 4.9). The isolated animal half-embryo (an1 + an2) forms a ciliated blastula, with an enlarged tuft, that does not gastrulate. Upon the addition of the veg1 tier (an1 + an2 + veg1), larvae lacking most of the gut and all of the skeleton are formed. If veg2 is added to an1 + an2, a more normal larva with gut and skeleton develops, but in this case the skeleton is derived from veg2 cells rather than from micromeres, as in normal development. Finally, when micromeres are recombined with an1 + an2, a fairly normal larva is formed with a skeleton derived from micromeres.

These experiments show that there is an animal-vegetal gradient of morphogenetic substances that is already present in the unfertilized egg and determines the fate of the cleavage cells along the animal-vegetal axis. The recombination experiments also show that the cells from the different tiers can reestablish the gradient by cellular interactions. Yet the molecular nature of these morphogenetic substances remains elusive. Even though sea urchin eggs and synchronized embryos can be obtained in large quantities, the biochemical approach to identifying morphogenetic substances has been unsuccessful. Only the genetic approach of identifying the genes encoding the morphogenetic substances and their subsequent cloning in organisms that are more amenable to genetics, such as *Drosophila,* have made it possible to isolate morphogenetic substances and to elucidate their mechanism of action.

Drosophila: *Determinants and Cell-Autonomous Differentiation*

Early *Drosophila* development (fig. 4.10) differs significantly from the development of nematodes, sea squirts, and sea urchins. First, the architecture of the egg is more elaborate, the three body axes—antero-posterior, dorso-ventral, and left-right—are already established during oogenesis, when the egg is formed. The second major difference concerns cleavage, which is delayed because the embryo develops first by nuclear divisions without subdivision of the cytoplasm. Only later do cell membranes, enclosing cortical cytoplasm, form around the preformed nuclei. After fertilization, the egg and

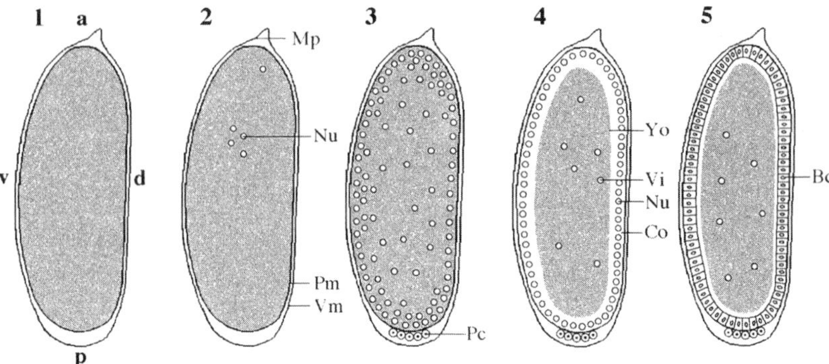

Figure 4.10
*Early embryogenesis of the fruit fly (*Drosophila melanogaster*). (1) In the egg the anterior-posterior (a, p) and the dorso-ventral (d, v) axes are already established. (2) Early cleavage. The sperm enters through the micropyle (Mp) and its nucleus fuses with the haploid egg nucleus to form the diploid zygote nucleus, which undergoes a rapid series of synchronous nuclear divisions in the egg cytoplasm, which is surrounded by the plasma membrane (Pm) of the egg. This multinucleated stage is called a syncytium. The syncytium is protected by two egg shells: the vitellin membrane (Vm) and the chorion (not shown). (3) At the 512-nuclei stage, some nuclei bud off from the posterior pole, become surrounded by a plasma membrane, and form pole cells (Pc), the primordial germ cells. (4) At the syncytial blastoderm stage, most nuclei (Nu) line up at the periphery of the egg in the cortical cytoplasm (Co), and only a few nuclei stay behind in the yolk (Yo) to form vitellophages (Vi), which are involved in the digestion of yolk. (5) Cellular blastoderm. The nuclei lined up in the cortex become separated by cell membranes (cellularization) and form a monolayer of blastoderm cells (Bc) that give rise to the somatic cells.*

sperm nuclei fuse, and the resulting zygote nucleus divides synchronously into 2 nuclei, then 4, 8, 16, and so on. These nuclear divisions are extremely rapid and last for nine minutes only at 25°C. After the ninth nuclear division, when 512 nuclei are formed, the nuclei begin to migrate from the center of the egg toward the periphery. Approximately a dozen nuclei bud off at the posterior pole and become enclosed by cell membranes. These pole cells are destined to become the future germ cells. The remaining nuclei, with the exception of a few that stay behind to become yolk cells, line up at the periphery of the egg cytoplasm. After the thirteenth nuclear division, cell membranes are assembled between the nuclei, leading to the formation of a single layer of cells, the cellular blastoderm.

What is the developmental fate of the pole cells, and how do they become determined? A variety of experiments indicate that the pole cells are the precursors of the germ cells that migrate from the posterior pole of the embryo to the gonads, where they later form eggs or sperm. As first shown by Rudolf Geigy, the pole cells can be ablated by ultraviolet (UV) irradiation of early embryos. The irradiated embryos give rise to sterile adult flies with empty gonads lacking germ cells. The same effect can be achieved by mutations that prevent pole cell formation. Such mutations have a maternal effect because they involve gene products that are supplied to the embryo by gene activity during oogenesis in the mother. Homozygous mutant mothers produce defective eggs that do not form pole cells. Such embryos can develop to adulthood, but the resulting adults lack germ cells and are therefore sterile. Such mutants are designated as *grandchildless,* since the homozygous mothers do produce children but do not have grandchildren, because their children lack germ cells.

An alternative way of studying the cell lineage of pole cells is to transplant single pole cells into the posterior pole of genetically marked host embryos of the same stage (fig. 4.11). Such recipients can produce progeny that is derived from the transplanted pole cell. The question of whether a pole cell is already determined—that is, committed to form germ cells—can be answered by heterotopic (to a different site) transplantation; if a genetically marked pole cell is transplanted from the posterior pole of the donor embryo to the midventral region of the recipient embryo, the transplanted pole cell may nevertheless find its way to the gonad and give rise to normal germ cells that can be distinguished from the recipient's germ cells by the marker gene. Because the transplanted cell differentiates according to its origin rather than its new location, this result indicates that the pole cells are determined to form germ cells at the time of their formation.

The next question that arises concerns the location and the nature of the determinants. Are the determinants localized in the cytoplasm at the posterior pole, or do the nuclei already carry germ cell determinants by the time they arrive at the posterior pole? These questions have been analyzed by cytoplasmic transplantation experiments. As in the case of *C. elegans,* the posterior polar cytoplasm contains specific structures, called polar granules, that are thought to be involved in germ cell determination. When cytoplasm is removed from the posterior pole of a donor egg with a micropipette and injected into the anterior pole region of a recipient egg, the immigrating anterior nuclei, instead of forming blastoderm cells, form cells that look like pole cells when examined by light or electron microscopy. But are they really functional pole cells? When these putative pole cells are transplanted from

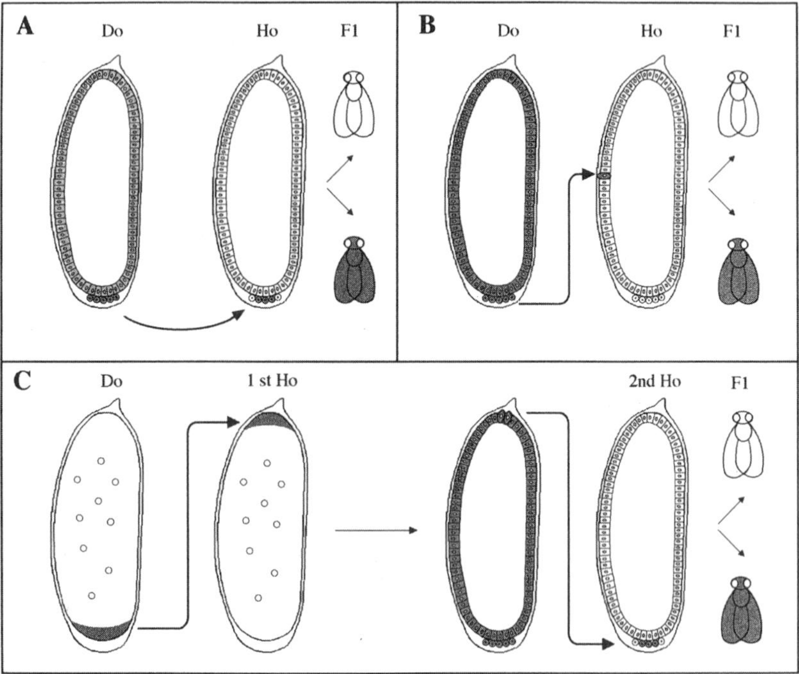

Figure 4.11
Germ cell determination in Drosophila: *(A) homotopic transplantation of genetically marked pole cells from the posterior pole of a black donor (Do) to the posterior pole of a white host (Ho) embryo; (B) heterotopic transplantation to the ventral side of the blastoderm host; (C) transplantation of posterior cytoplasm to the anterior pole of the first host; induction of anterior pole cells in the first host (black) and their transplantation from the posterior pole to a different site on the second host (white), where they can give rise to black progeny. Modified from R. Wehner and W. J. Gehring,* Zoologie, *23d ed. (Stuttgart: Thieme Verlag, 1995), fig. 3.24.*

their anterior position back to the posterior pole of a genetically marked recipient embryo, they can in rare instances give rise to germ cells that produce progeny indicating that they are functional germ cell precursors. The reasons why they are rarely functional are not understood, but this result has been obtained independently in two laboratories. There may be competition between the injected germ cell determinants and the resident anterior somatic determinants, which is discussed later. In conclusion, these transplantation experiments strongly suggest the presence of germ cell determinants localized in the posterior polar cytoplasm, but their molecular nature remains a challenge for future molecular genetic experiments.

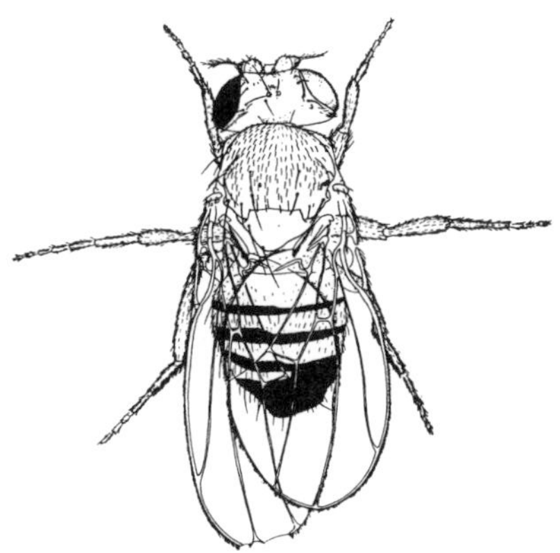

Figure 4.12
Gynandromorph. The cells on the left side of this fly carry two X-chromosomes and are female (abdominal pigmentation), expressing the dominant marker genes white$^+$ *(red eye) and* Notch *(notched wing), whereas the cells on the right side have lost one X-chromosome (XO), exposing the recessive marker genes* white *(white eye) and* forked *(bristles), and are male with black abdominal pigmentation. From T. H. Morgan, C. B. Bridges, and A. H. Sturtevant,* The Genetics of Drosophila *(New York: Garland, 1988).*

As early as 1929, Alfred Sturtevant succeeded in analyzing the lineage and fate of the nuclei before blastoderm formation by means of gynandromorphs. Gynandromorphs (from Greek *gyne,* female, and *andros,* male) are genetic mosaic animals consisting of both male and female cells (fig. 4.12). Such individuals are also found exceptionally in human populations. In *Drosophila,* gynandromorphs arise by the loss of one of the two X-chromosomes in female (XX) embryos, because cells with a single X-chromosome are male. In certain mutant stocks, a particular X-chromosome is frequently lost from both daughter nuclei during the second nuclear division, giving rise to a mosaic embryo with two XX and two XO nuclei. Such embryos develop into flies consisting of approximately 50 percent female (XX) and 50 percent male (XO) cells. Examination of individual blastoderm embryos indicates that the two cell types (XO and XX) are randomly distributed in the blastoderm (fig. 4.13). Even though the nuclei do not intermix freely, the division line between XX and XO cells is random. The path of migration of the nuclei to the egg surface is thus not predetermined but random.

The later fate of the blastoderm cells can be followed by marker genes. By introducing recessive marker mutations on the normal X-chromosome that is retained in the male cells, the XO male cells can be distinguished from the XX female cells. For example, the recessive yellow (y) mutation can be introduced into the normal X-chromosome. Because the second X-chromosome in the female carries the wild-type gene (y^+), which is dominant over the yellow mutant gene (y), the female cells (y^+/y) express the wild-type body color, which is brown. However, the male cells carrying only one mutant X-chromosome (y/O) express the yellow body color. By looking at the distribution of brown (XX) and yellow (XO) cells in a large set of gynandromorphs (see fig. 4.13), it becomes obvious that it is random. Because the adult fly is formed essentially by an expansion of the blastoderm, the migration of the nuclei to the egg surface must also be random. We can therefore conclude that the lineage of the nuclei before blastoderm formation is irrelevant for their developmental fate. Rather, the cortical cytoplasm must contain the positional information that instructs the nuclei. A nucleus arriving at the anterior pole, for example, is instructed to form an anterior structure (the molecular basis of this instruction by anterior cytoplasmic determinants is discussed in the next chapter). From these gynandromorph studies a second and equally important conclusion can be drawn: the nuclei are equivalent

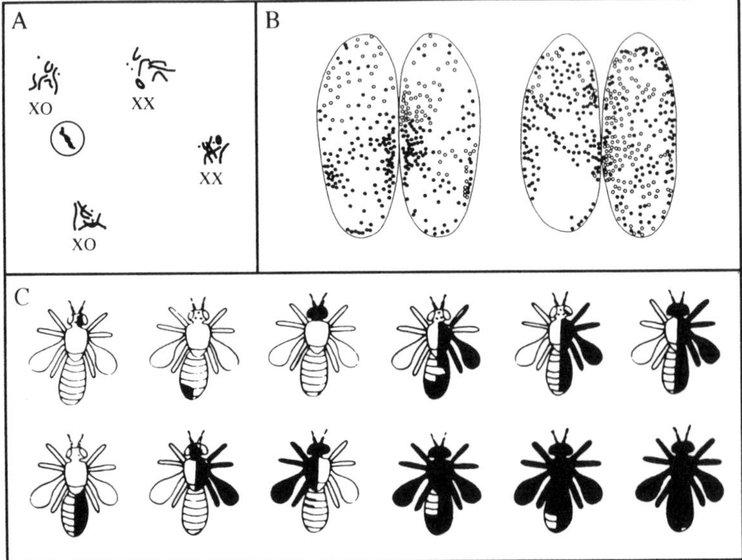

Figure 4.13
Gynandromorphism caused by X-chromosome loss: (A) loss of the two interlocked ring-shaped X-chromosomes (encircled) from two of the four nuclei during the second nuclear division, leading to two male (XO) and two female (XX) nuclei; (B) distribution of female (white) and male (black) nuclei on the surface of the syncytial blastoderm embryo, which is cut open to show the two halves; (C) distribution of female (white) and male (black) tissue on a set of 12 gynandromorphs. From M. Zalokar, I. Erk, and P. Santamaria, Distribution of ring-X chromosomes in the blastoderm of gynandromorphic D. melanogaster, Cell *19 (1980): 131–41.*

and totipotent before the formation of the blastoderm; if they are incorporated into a pole cell, they become future germ cells, if they arrive at the anterior pole they will form a head structure, and so on. They thus receive their positional information from the cortical cytoplasm. This conclusion has been confirmed by nuclear transplantation experiments and isolation of the cytoplasmic determinants.

The fate and determination of the blastoderm cells has been analyzed by various genetic and embryological experiments. In contrast to the nematode *C. elegans* or ascidians, the cell lineage of blastoderm cells in *Drosophila* is not fixed but variable. If single blastoderm cells are genetically marked, for example by the yellow marker or by injection of a marker dye, the resulting clones of marked cells do not always contain the same number of cells, nor

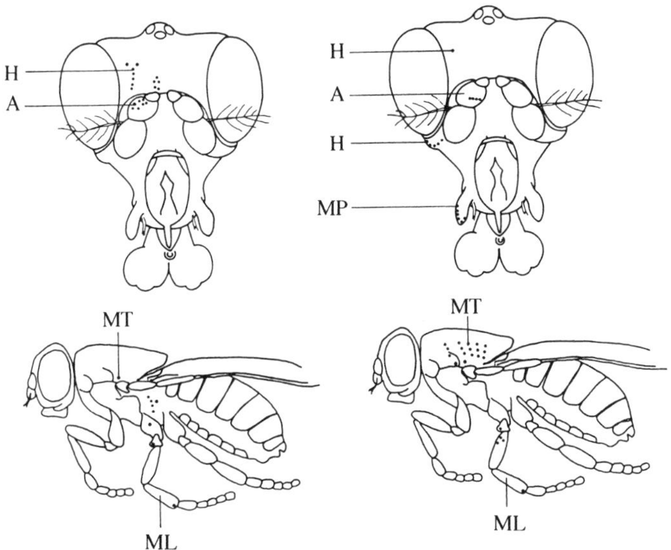

Figure 4.14
Patterns of clones of genetically marked bristle cells (dots) on the head and thorax of the fruit fly. The mother cells of these clones were marked at the blastoderm stage. Their descendants are variable in number and occupy varying positions on the surface of the fly that may extend between the areas derived from two imaginal discs: the eye and antenna disc (upper row) and the wing and middle leg disc (lower row). Abbreviations: H *head,* A *antenna,* MP *maxillary palpus,* MT *mesothorax,* ML *middle (mesothoracic) leg. Courtesy of E. Wieschaus.*

do they occupy fixed positions at later stages of development. The clones are variable in size and topography (fig. 4.14). There are restrictions, however; the clones are confined to a given body segment, for example, the first or the second thoracic segment, but they do not overlap between segments. This restriction may indicate that the first steps of determination do take place during blastoderm formation. Indeed, if anterior half blastoderms are dissociated into single cells and cultured, they can only produce structures derived from the anterior head and the thoracic segment even if intermixed with marked posterior blastoderm cells. Thus, anterior cells are determined to form anterior structures and develop autonomously.

The posterior cells cannot induce anterior cells to change fate and to differentiate into posterior structures and vice versa. This result has been confirmed by transplantation of single cells to the same or a different site of a recipient blastoderm. If a cell from the anterior pole of the blastoderm is transplanted to the same position of a genetically marked recipient blastoderm, the donor cell associates with the resident cells, becomes integrated, and forms part of the head tissue. In contrast, if an anterior cell is transplanted to a posterior site, it cannot integrate into the surrounding tissue, but it may form some head structures in the cavity of the abdomen of the recipient fly. This shows clearly that the donor blastoderm cell is already determined or committed to form head structures and that it develops independently of the surrounding cells.

Amphibians: Spemann's Organizer and the Role of the Genes

Newts and frogs have relatively large eggs, one to two millimeters in diameter, and their embryos are easily operated on. For this reason, many fundamental embryological experiments have been carried out in these amphibians. Because amphibians are vertebrates like ourselves, the findings made in amphibian embryos may in many cases also apply to humans, but as we shall see later, this argument may also apply to flies, much more so than was once thought. In the present context I describe two fundamental experiments carried out first in amphibians, Hans Spemann's organizer experiment and nuclear transplantation.

In his delightful book *The Heritage of Experimental Embryology: Hans Spemann and the Organizer,* Viktor Hamburger vividly describes the history of the organizer experiment and places it in historical context. Not only was Hamburger a graduate student of Spemann, and therefore an insider, but he has made outstanding contributions to experimental embryology himself, so that he is in an excellent position to tell the story of this important discovery. Hans Spemann himself wrote *Embryonic Development and Induction* (Yale University Press, 1936), a synthesis of his work based on the Silliman Lectures he gave at Yale University in 1931.*

As a prelude we first have to know how the amphibian embryo develops with special emphasis on gastrulation. Before fertilization, the spherical egg is polarized into a pigmented animal hemisphere and a paler vegetal hemisphere loaded with heavy yolk granules; the egg nucleus, visible as a pale area, is located near the animal pole. The pigmentation of the frog (*Rana*) egg permits one to see the cytoplasmic rearrangements that occur upon sperm entry. The black pigment granules are confined to a superficial layer that covers the animal hemisphere like a cap. When the sperm penetrates, the pigment cap shifts by approximately twenty degrees relative to the interior cytoplasm to an oblique position and exposes the gray interior cytoplasm on the opposite side of where the sperm has entered. The grayish area is called the gray crescent. This movement of the peripheral (cortical) cytoplasm relative to the interior cytoplasm is indispensable for later gastrulation movements. The first cleavage furrow begins to form at the animal pole and proceeds meridionally toward the vegetal pole. If it bisects the gray crescent cytoplasm so that both blastomeres receive roughly equal amounts, the isolated blastomeres can each give rise to a complete tadpole and form identical twins. If, however, one of the blastomeres does not contain enough gray crescent cytoplasm, it fails to gastrulate when isolated. In amphibians the first two blastomeres are large enough to be separated by ligation with a fine hair from a child. Ligation can also produce twins at later stages until gastrulation, when determination takes place. Gastrulation involves the movement of cell layers resulting in the separation of the three germ layers and the basic organization of the body plan.

Cleavage of the early embryo causes the formation of a hollow sphere of cells, the blastula, which corresponds to the blastoderm stage as described for *Drosophila.* At the animal pole are the pigmented cells that will give rise to the outer germ layer (ectoderm) forming the skin and the nervous system. At the vegetal pole are the yolk-laden cells that will form the endoderm, which differentiates into the gut. The intermediate cells form the middle germ layer, the mesoderm. Gastrulation is the process of invagination of the

*Spemann's book was one of my incentives in trying to write up the Terry Lectures I gave at Yale—my second alma mater—in 1993, trying to stay in this tradition.

vegetal hemisphere through the blastopore into the interior, while the dorsal hemisphere stretches and flattens to cover the invaginated parts (fig. 4.15).

The changes that take place during invagination are very complex and can be dealt with here only in broad outline. While invagination proceeds the invaginating material splits: the vegetal part, composed of yolk-laden

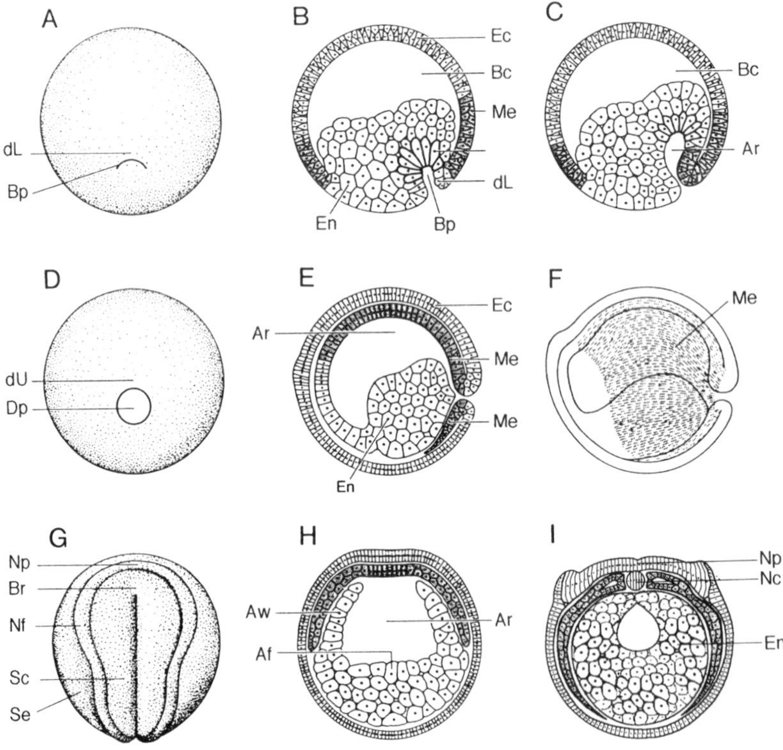

Figure 4.15
Gastrulation and neurulation in amphibians: (A) surface view of young gastrula stage embryo showing the blastopore and the dorsal lip (Spemann's Organizer); (B) section through (A) illustrating the invagination of flask-shaped cells through the blastopore; (C) section through midgastrula-stage embryo; (D) surface view of late gastrula with round blastopore (yolk plug); (E) section through (D); (F) illustration of how the mesoderm invaginates and extends between the endoderm and ectoderm; (G) surface view of the neurula stage; (H) frontal section indicating the upward growth of the archenteron walls; (I) dorsal closure of the archenteron and differentiation of the notochord. Abbreviations: Af *archenteron floor,* Ar *archenteron,* Aw *archenteron wall,* Bc *blastocoel,* Bp *blastopore,* Br *brain area,* Dp *yolk plug,* dU *dorsal lip,* Ec *ectoderm,* En *endoderm,* Me *mesoderm,* Nc *notochord,* Nf *neural fold,* Np *neural plate,* Sc *spinal chord area,* Se *skin ectoderm. From R. Wehner and W. J. Gehring,* Zoologie, *23d ed. (Stuttgart: Thieme Verlag, 1995), fig. 3.17.*

cells, forms a trough, the precursor of the gut, called the archenteron. The lateral walls of the trough grow upward, split from the more dorsal cell layer to form the archenteron roof, and eventually fuse in the dorsal midline, forming a hollow tube, the future intestine. The archenteron roof subsequently forms the mesodermal structures, including the notochord in the dorsal midline, two rows of somites (muscle segments), one on either side of the notochord, and the lateral plates. In a subsequent process called neurulation, the external cell layer (ectoderm) located above the archenteron roof forms the neural plate, which subsequently rolls up into the neural tube. The anterior part of the neural tube eventually gives rise to the brain, and the posterior part gives rise to the spinal chord. At the end of this process, the embryo elongates to form the tailbud stage at which the basic organization of the embryo is clearly visible (fig. 4.16). In a cross-section we can recognize the basic body plan of a vertebrate embryo, which is basically the same in fish, amphibians, reptiles, birds, and mammals. The tailbud embryo reflects another important fact: the timing of development; clearly, the head differentiates first and the more posterior structures gradually differentiate later, the tail still being an undifferentiated bud.

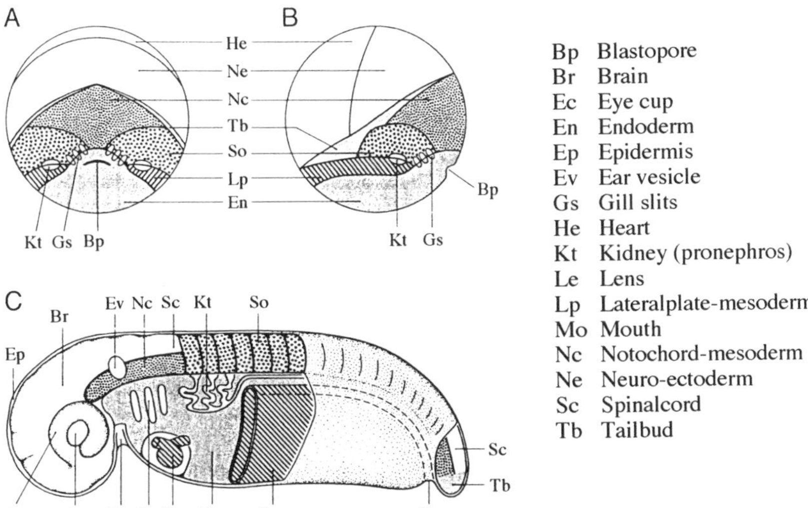

Figure 4.16
Fate map and organization of the amphibian embryo: (A) fate map, dorsal view; (B) fate map, lateral view; (C) organization of the embryo. From R. Wehner and W. J. Gehring, Zoologie, *23d ed. (Stuttgart: Thieme Verlag, 1995), fig. 3.18.*

Early experiments by Hans Spemann had shown that the capacity of separated half-embryos to produce duplications ceases at the end of gastrulation. In order to find out whether at this stage the various parts of the embryo become committed to a certain developmental pathway, Spemann wanted to make a comprehensive test of their state of determination by transplanting various parts from an early donor gastrula to different positions of a recipient embryo of the same stage. If such transplants developed autonomously, the experiment would indicate that the transplanted cells were already committed. If, however, the cells adopted the fate of the surrounding recipient cells, they would be considered to be undetermined at this stage.

At the early gastrula stage, parts destined to become the neural plate can be exchanged freely with future epidermis, and the cells differentiate in accordance with their new environment, indicating that they are not yet determined. In contrast, if the parts are transplanted right after gastrulation, when the neural plate just becomes visible, the cells differentiate autonomously according to their origin and not according to their new environment. If, for example, part of the neural plate containing the area of the future eye is transplanted to the flank of the recipient embryo, an additional eye is formed instead of epidermis in the flank of the recipient. This experiment indicates that determination of the eye primordium takes place during gastrulation. Similarly, a piece of epidermis differentiates autonomously when transplanted to the neural plate after gastrulation.

The only part that behaved differently was the dorsal lip of the blastopore. When Spemann transplanted the dorsal lip from the donor into the flank of the recipient at the early gastrula stage, the transplant did not differentiate according to its new environment but rather invaginated into the interior, forming a second neural tube underlain by a notochord and two rows of somites. Spemann's early experiments suffered from a serious limitation, however. Donor and recipient were taken from the same newt species, *Triturus taeniatus*, the common newt, and could be distinguished only by the varying degrees of pigmentation of the cells. This made it difficult to assign certain cells unequivocally to the donor or the recipient. At this time, however, the leading American embryologist Ross Harrison had introduced the method of heteroplastic transplantation, which involves two different animal species showing marked differences in cell morphology. At these early developmental stages there is no immunological rejection of the graft by the recipient. Spemann thus switched to heteroplastic transplantation and assigned the crucial experiment to his student Hilde Mangold. As donor, the almost unpigmented gastrula of the crested newt, *Triturus cristatus,* was used and the dorsal blastopore lips were transplanted to the left flank of a recipient gastrula of the common newt (fig. 4.17). Rather than forming epidermis corresponding to its new environment, the implant almost completely invaginated and induced the formation of a secondary neural plate in the host cells. The donor cells also induced secondary somites on either side of a secondary notochord. The notochord was entirely of donor origin, as indicated by the lack of pigmentation, and the somites were composed of both host and donor cells. The most important case, Um 132, is illustrated in figure 4.17, and the seminal paper by Spemann and Mangold (1924) describes only four other experimental cases, but at that time experimental embryology was not a quantitative science and the results were neither tabulated nor

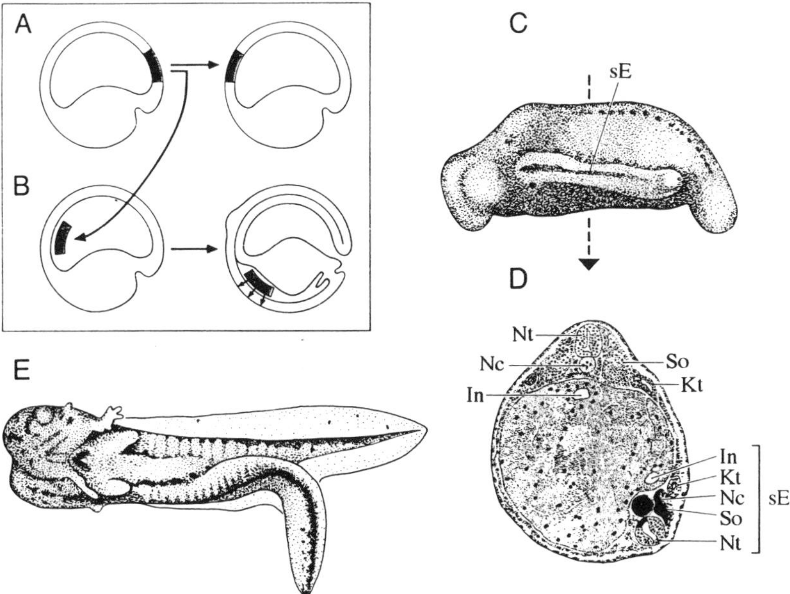

Figure 4.17
Organizer experiment. (A) Transplantation of the dorsal blastopore lip (organizer) to the ventral side of a recipient gastrula. (B) "Einsteckversuch": insertion of the dorsal lip into the blastocoel. (C) Induction of a secondary embryo (sE) on the ventral side. (D) Cross-section of embryo in (C). Donor tissue in black. The notochord of the secondary embryo is derived from donor cells. The neural tube and one somite are composed of both donor and recipient cells. (E) Development of a secondary embryo. From R. Wehner and W. J. Gehring, Zoologie, *23d ed. (Stuttgart: Thieme Verlag, 1995), fig. 3.27.*

analyzed statistically. Nevertheless, these results have been confirmed many times by other authors, and the skillful students of my embryology course were able to reproduce the experiment consistently. Johannes Holtfreter simplified the procedure by inserting the blastopore lip directly into the cavity (blastocoel) of the recipient embryo opposite to the blastopore and succeeded in obtaining essentially complete secondary embryos (fig. 4.17). Because the blastopore lip is capable of inducing the formation of a new embryonic axis and of organizing a complete embryo, the blastopore lip was designated as organization center, or organizer.

Spemann interpreted the organizer merely in epigenetic terms that emphasized tissue interactions. He did not realize that a large number of genes had to be activated to induce the formation of a secondary embryo. He did not see the straightforward connection between embryology and genetics, even though his collaborator Oscar Schotté succeeded in a spectacular experiment that convincingly illustrated the importance of the genes. As Viktor Hamburger reports in his book *The Heritage of Experimental Embryology*, Oscar Schotté joined Spemann in 1928 and soon began a dream experiment that at the time seemed completely unrealistic. Schotté transplanted ventral ectoderm of the frog gastrula to the mouth region of a newt gastrula and vice versa. When discussing the experiment with Viktor Hamburger, Schotté pledged that he would not shave until the experiment was a success; his beard grew and grew, until one morning he appeared on the scene, clean-shaven and smiling: the first transplanted larva had survived. The larva was a salamander with the mouth parts of a frog. The frog cells had responded to the signal provided by the salamander recipient by forming a mouth, but the structure of the mouth parts was obviously determined by the frog genes, so that the frog's horny jaws and suckers were formed. Spemann, however, continued to emphasize the interaction between the embryonic tissues, which are not species-specific; the activation of the frog genes in the responding tissue did not figure prominently in his thoughts. The idea that this method could be used to study gene activation was mentioned neither in Spemann and Schotté's paper of 1932 nor in Spemann's 1936 book based on his Silliman Lectures.

The study of the organizer took a new turn when Johannes Holtfreter showed that even dead embryonic tissue can induce a neural plate, which argued strongly in favor of the chemical nature of the inducing agent. Yet the nature of the inducer that emanated from the organizer tissue remained elusive for more than seventy years. The biochemical approach was unsuccessful. Embryological studies have since shown that neural induction is preceded by mesoderm induction: when animal pole cells are explanted alone, they form ectodermal structures only, whereas vegetal pole cells forms endoderm only. However, animal pole cells can be induced to form mesodermal structures if they are recombined with vegetal pole cells (see the proposed model, fig. 4.18). The organizer, localized in the mesoderm, is then thought to signal to the overlaying ectoderm and induce the formation of the neural

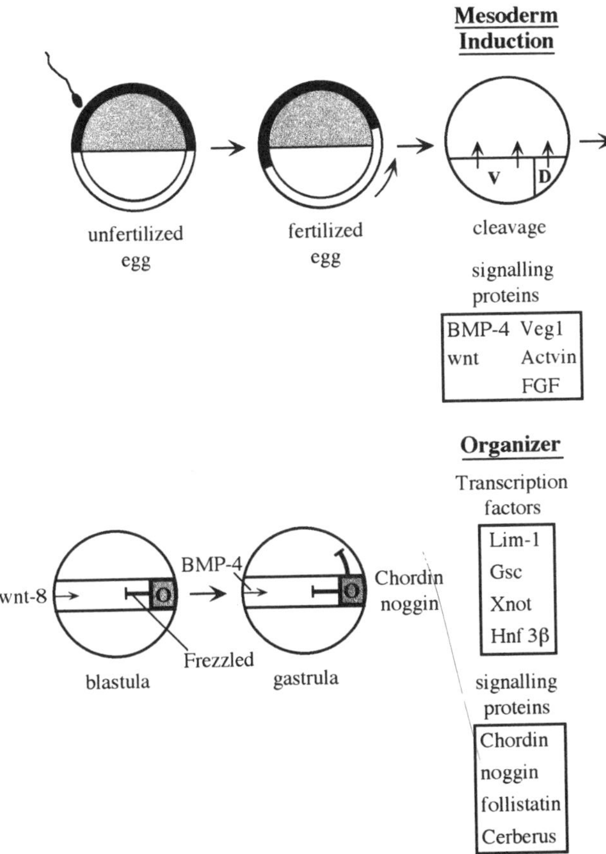

Figure 4.18
Mesoderm induction and organizer action. Upon sperm entry the cortical egg cytoplasm (black) rotates and exposes the future organizer region (o). At least five genes (box) encoding signaling proteins are involved in mesoderm induction. Several genes encoding transcription factors and signaling proteins are specifically expressed in the organizer region. Their protein products interact with bone morphogenetic protein 4 (BMP-4) in the induction of the various mesodermal tissues and the neural plate. Arrows indicate activation, T-bars repression.

plate, which is called neural induction. Activin, a putative mesoderm-inducing factor, was isolated biochemically from *Xenopus* tissue culture cells by James Smith. However, these biochemical experiments must be complemented with genetic experiments in order to demonstrate that activin is indeed a natural mesoderm-inducing factor in the embryo. *Xenopus* and other amphibians are poorly suited for a genetic approach, and so progress has been slow as compared to that in *Drosophila.* Using reverse genetic procedures—that is, starting from cloned isolated genes—a number of putative signaling and transcription factors specific for the organizer have been identified, and a clear picture is beginning to emerge. The current view of organizer formation and action implies the following steps: (1) induction of the mesoderm by signaling molecules encoded by such genes as activin, *vg1,* and *wnt;* (2) expression of homeobox genes specifically in the area of the dorsal blastopore lip such as *goosecoid, lim-1, Xnot,* and another transcription factor of the forkhead family (HNF3b); and (3) activation by these regulatory genes of the signaling molecules that are secreted, such as chordin, noggin, and follistatin. These signaling molecules are thought to counteract BMP4, another signaling molecule, by binding to it and blocking the interaction of BMP4 with its receptors in the ectoderm. The inhibition of the ventralizing signals of BMP4 leads to the differentiation of the nervous system on the dorsal side of the embryo. Because *Xenopus* is difficult to manipulate genetically, the respective mouse genes have been cloned. Using homologous recombination as a tool, the respective genes can be inactivated in the mouse embryo. A "knock-out" mouse in which the *lim-1* homeobox gene has been inactivated has a dramatic effect: it develops essentially without a head. It is unclear why the nervous system of the trunk is not affected, but the possibility that there is genetic redundancy has to be seriously considered because *lim-1* is also expressed in the trunk. By making double mutant knock-out mice this question can be answered and the mechanism of neural induction may finally be elucidated using these genetic techniques.

The first genetic theory of development was proposed by August Weismann, who was a strong proponent of Darwinism and held the chair of zoology at the University of Freiburg in Germany. In *The Germ Plasm: A Theory of Heredity,* published in 1892, Weismann proposed that the hereditary material in the egg nucleus is split during cleavage by a defined sequence of unequal nuclear divisions. In this way the genes that he called hereditary determinants would be segregated into different cell lineages and would determine the specific differentiation of those cells in which they resided. He also discussed the alternative in which the daughter cells of the egg and their descendants would receive the entire hereditary material by equal nuclear divisions. In this case, cellular differentiation would be determined by interactions between nucleus and cytoplasm and between cells and their neighbors leading to differential gene activation. Weismann opted for the first hypothesis, which proved to be wrong. Driesch's experiment separating the individual cells at the two- and four-cell stages in sea urchins argued against Weismann's theory because each of these cells could develop into a complete larva. At least up to the four-cell stage, therefore, each cell contains a complete set of genes.

It took many years, however, to prove that cells at later stages of development also contain a complete genome. This evidence came from nuclear transplantation experiments. When I was a student in Ernst Hadorn's laboratory, Fritz Baltzer came to Zurich to give a seminar on nuclear transplantation experiments carried out by Thomas Briggs and Robert King in the 1950s. Fritz Baltzer was, like Hans Spemann, a former student of Theodor Boveri in Würzburg. In 1918, when Spemann succeeded Weismann at the University of Freiburg, he asked Baltzer to join him and Baltzer accepted a professorship. In 1921 he returned to his home country, Switzerland, to take the chair of zoology in Bern, where he became the leading developmental biologist. The most prominent of Baltzer's students was my mentor, Ernst Hadorn. Baltzer and Hadorn shared a deep interest in the role of the genome in development and both had worked on merogones, organisms that develop from enucleated eggs with the paternal genome only provided by the sperm. Such merogones have only a limited developmental potential, in part because they are haploid and in part because they lack maternal chromosomes. I felt privileged to meet the grand old man, who was strong like a bear with short, curly hair and a mustache. He presented the nuclear transplantation experiments with great vigor and discussed them critically in the light of previous experiments. Hadorn had great respect for his mentor and he would never call him by his first name. He always called him Herr Baltzer, though leaving out "Herr Professor."

To my surprise, Hadorn seemed to be more conservative than his mentor, and he had a hard time believing that Briggs and King had succeeded in an experiment that he had tried without success in newts. Briggs and King had removed the nucleus of a frog egg mechanically and injected a nucleus taken directly from a cell of a cleavage stage or blastula embryo with a micropipette. They found that such eggs could undergo cleavage and in a fraction of the cases even become normal tadpoles. This indicated that the cells contain a complete genome, at least up to the blastula stage, thus defeating Weismann's theory. However, controversy continued for many years whether completely differentiated cells retain a complete genome because of a decline in the success rate as older and older cells were taken as source of the donor nuclei. The controversy was settled by John Gurdon, who showed that fertile frogs could be obtained from nuclei of fully differentiated intestinal tadpole cells from the clawed frog *Xenopus*. Some doubts lingered, however, because the nucleus of *Xenopus* eggs was inactivated by ultraviolet irradiation rather than removed completely and because the success rate was below 1 percent. In addition, mammals differ from frogs in that they exhibit the phenomenon of genomic imprinting in both the egg and sperm nucleus. Nevertheless, all the gene cloning experiments indicate that with the exception of a few special cases, the genome does not undergo irreversible changes in the course of development and that somatic cells generally contain a complete genome. These findings support the alternative hypothesis that was rejected by Weismann and involves differential gene activity as the basis of cell differentiation.

Mammals: The American Way of Life

The mammalian embryo represents the best example for the American way of developing, as Sidney Brenner has noted. The ancestry of a cell does not influence development; what is important are the interactions with neighboring cells. Mammalian embryo cells are very flexible, they move around to find their proper place, and if they die, other cells take over their function. This method of development is apparent from blastomere isolation experiments in mice: separation of the blastomeres at the two- or four-cell stage produces identical twins and quadruplets, respectively, and if two embryos are fused at the eight-cell stage, they form a mosaic (or chimeric) mouse of normal size. The cell number is adjusted to the normal value before gastrulation in a mysterious way that is not understood.

If blastomeres of a black embryo are combined with blastomeres of an albino mutant, the resulting mosaic animal shows an irregular pattern of black and white areas (fig. 4.19). This is because the neural crest cells leave the dorsal midline and migrate to either side of the animal, giving rise to pigment cells at the base of the hairs. The clones of neural crest cells do not occupy defined areas of the skin as they would, for example, in sea squirts (see fig. 4.6), but their migratory path is variable and so there is intermixing with cells from other clones.

Figure 4.19
A white mouse foster mother with her progeny, some black and some chimeric (black and white). Courtesy of K. Bürki.

After the sixteen-cell stage, the cells differentiate into the trophectoderm cells, which later contribute to the placenta, and into the inner cell mass, which gives rise to the embryo proper. The trophectoderm cells are irreversibly committed, but the cells of the inner cell mass remain totipotent in the sense that they can give rise to any part of the future embryo, including the germline. Such a cell can be transplanted from the inner cell mass of one blastocyst embryo into the cavity of a genetically marked blastocyst of the same stage and give rise to as much as 90 percent of all cells of the resulting mouse. This fact prompted Martin Evans and Gail Martin to establish cell lines from the inner cell mass of such blastocysts, which retained their totipotency in culture. These embryonic stem cells were crucial to the development of gene transfer techniques and were a breakthrough in mouse developmental genetics. Before then, the general status of the mouse field had been characterized in the *Laboratory Manual for Manipulating the Mouse Embryo* by the sobering statement: "It is hard to escape the general conclusion that in spite of all the work that has gone into the phenotypic analysis and breeding of mouse mutants over the years, very little has so far been learnt from them about the genetic control of differentiation and morphogenesis." The new efficient gene replacement methods, resulting largely from the pioneering work of Mario Capecchi, are outlined in figure 4.20. They are based on the fact that mammalian cells have efficient enzyme systems that

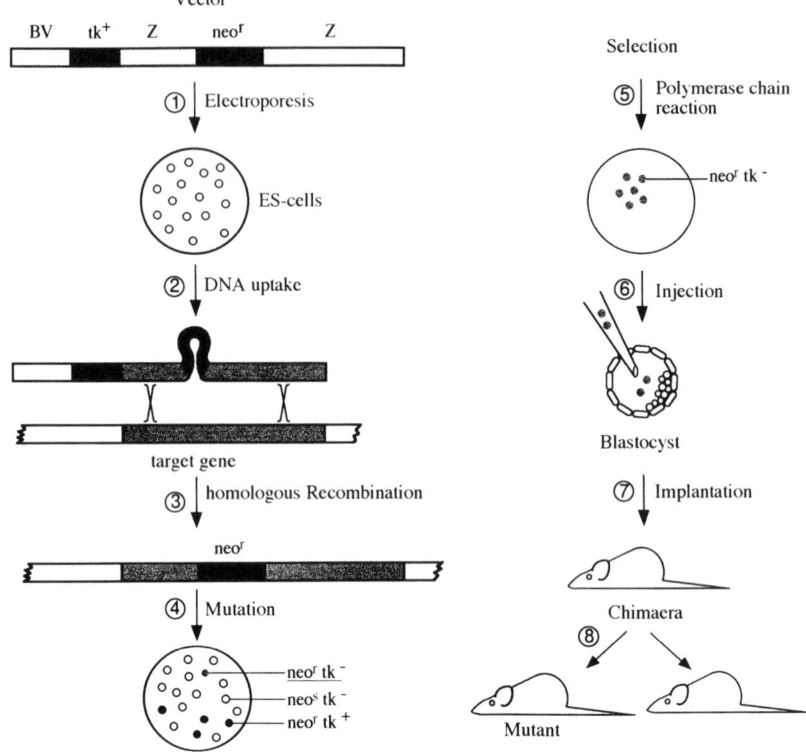

Figure 4.20

Gene targeting in mice. (1) Two selective marker genes, thymidine kinase (+k⁺) and neomycine-resistance (neo⁺), are inserted into the cloned gene Z and amplified in a bacterial vector (BV). The DNA is added to cultured embryonic stem cells (ES-cells), which are subjected to electrical fields that induce pore formation in the cell membrane (electroporation). (2) The DNA is taken up by the ES-cells through the pores. (3) In a process called homologous recombination, the transferred DNA molecule can associate with the resident target gene and replace the resident gene as a consequence of two crossover events (X) between homologous sequences. (4) As a result of homologous recombination, the resident gene is substituted by the transferred gene carrying the neomycine marker, which interrupts the coding region of the target gene and thereby inactivates it. (5) Cells carrying such mutations are neomycine resistant, have lost the tk gene (tk⁻), and can be isolated on a selective medium. The presence of the mutation in the target gene can be confirmed by the polymerase chain reaction. (6) neo^R tk⁻ cells can be grown in culture and injected into recipient blastocysts carrying different genetic markers. (7) Upon implantation into a foster mother, chimeric progeny consisting of neo^R tk⁻ and recipient cells can be obtained. (8) Mutant mice can be generated in the next generation if the chimeric parent produces mutant germ cells. From R. Wehner and W. J. Gehring, Zoologie, *23d ed. (Stuttgart: Thieme Verlag, 1995), box 2.2.*

allow them to recognize DNA molecules with identical base sequences and to integrate them into the chromosomes at corresponding positions by a process called homologous recombination. An engineered DNA molecule containing selectable marker genes and a designed mutation flanked by extended regions of sequence homology can be introduced into embryonic stem cells. Recombinant cells in which the resident gene is replaced by the designed gene can be selected in culture and a clone of recombinant cells can be isolated. The recombinant cells can be injected into a suitable recipient blastocyst and transferred into a foster mother that can give birth to a chimeric mouse. If the recombinant cells have colonized the gonad, the chimeric mouse can be bred, and in the next generation, mutant mice can be obtained that carry the replaced gene in all their cells. This method has revolutionized mouse developmental genetics.

Nevertheless, mammalian developmental genetics will never proceed as rapidly as the genetic study of *Drosophila* or *Caenorhabditis.* Not only is the mammalian genome much larger, but the generation time is much greater, and the embryos develop much more slowly. In addition, mammalian embryos grow in the protective and nutritive environment of the mother, making it harder to handle them. For these reasons a sizable group of developmental geneticists have recently switched to another vertebrate embryo, the zebra fish (*Danio rerio*). This fish, introduced by George Streisinger, has a beautifully transparent embryo that develops outside the mother, has a short generation time, produces a large number of progeny, and lends itself much better for genetic and developmental analysis. A substantial collection of mutants has been isolated. The zebra fish has drawbacks, however—the lack of an efficient method for gene transfer and the duplication of parts of its genome. Nevertheless, I expect that the zebra fish will contribute significantly to our understanding of vertebrate development.

This guided tour through embryology has demonstrated the great variety of modes of development ranging from worms with a fixed cell lineage to mammals displaying great flexibility, but there is a common underlying principle: the genetic control of development. DNA contains a detailed developmental program that has been deciphered most thoroughly in *Drosophila.* In the following chapters we shall explore this program in the developing fruit fly from the egg to metamorphosis and finally consider its evolutionary aspects.

Figure 5.1
The symbolic egg on the roof of Salvador Dali's villa in Port Lligat, Spain. Photograph by the author.

~

5
The Secret Is in the Egg

When you come over the hill from Cadaquez on the Catalonian coast to the small fishing village of Port Lligat, a gigantic white egg appears above the silver-green olive trees that surround the villa of Salvador Dali (fig. 5.1). This is where Dali spent most of his artistic life with his muse, Gala. Starting with a fisherman's cottage with no electricity or running water, over time Dali created a fantastic villa whose roof is decorated with two giant heads inclined toward each other, a sacred Indian elephant, and two gigantic eggs visible among the treetops from a distance. Dali was fascinated by the egg as a symbol for the origin of life, and he devoted several works of art to it. Eggs and early cleavage-stage embryos also decorate the roof of Dali's museum in Figueras (fig. 5.2).

It is sobering to realize that we have all arisen from a tiny egg cell about a hundred micrometers in diameter, barely visible to the naked eye. No matter how important we become later in life as politicians, managers, artists, scientists, or rock singers, we all began as a tiny little egg. On fertilization all the information required for future development is contained in that egg, and since the time of Aristotle, biologists have tried to unravel its secrets.

In addition to genetic information, the fertilized egg contains in its cytoplasm spatial information relating to the architecture of the future embryo. Most egg cells are polarized and specify at least one of the future axes of the embryo. Most isolated cells are not spherical or droplike as an amoeba but acquire a defined shape. This is due to their cytoskeleton, which may either be on the outside as a cell wall or shell or on the inside as structural filaments or tubules. In addition, microfilaments and microtubules can move molecules, higher-order aggregates, and organelles around inside the cell in a directed way, thereby maintaining cellular architecture and providing flexibility. Only a few cases of apolar, spherically symmetrical eggs, such as that of kelp, *Fucus*, are known. In such cases the egg is polarized by an external stimulus, such as illumination in the *Fucus* egg.

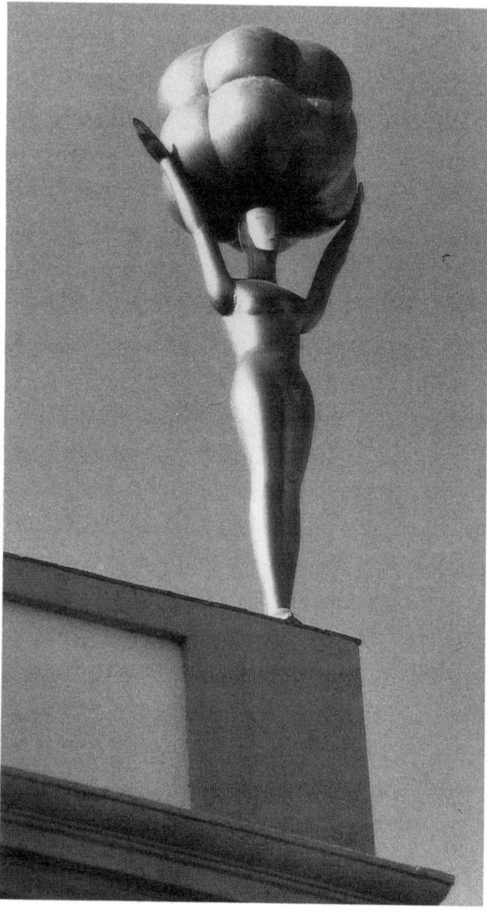

Figure 5.2
The cleaving embryo that surmounts the roof of the Dali Museum in Figueras, Spain. Photograph by the author.

In sea squirts, the egg is polarized in the ovary, where the animal-vegetal axis is laid down, but sperm entry causes extensive cytoplasmic movements that lead to bilateral symmetry. In addition to the nucleus, the sperm provides two centrioles to the egg that serve as organization centers for the microtubules that are involved in the movement of the sperm nucleus toward the center of the egg, where it unites with the egg nucleus. Microfilaments and microtubules are also involved in the cytoplasmic movements that are reflected by the formation of the yellow cap and the yellow crescent (pl. 2).

In *Drosophila* all three major body axes, antero-posterior, dorso-ventral, and left-right, are built into the egg during oogenesis and reflected by the shape of the egg shells (fig. 5.3). The oocyte is polarized as it develops in the

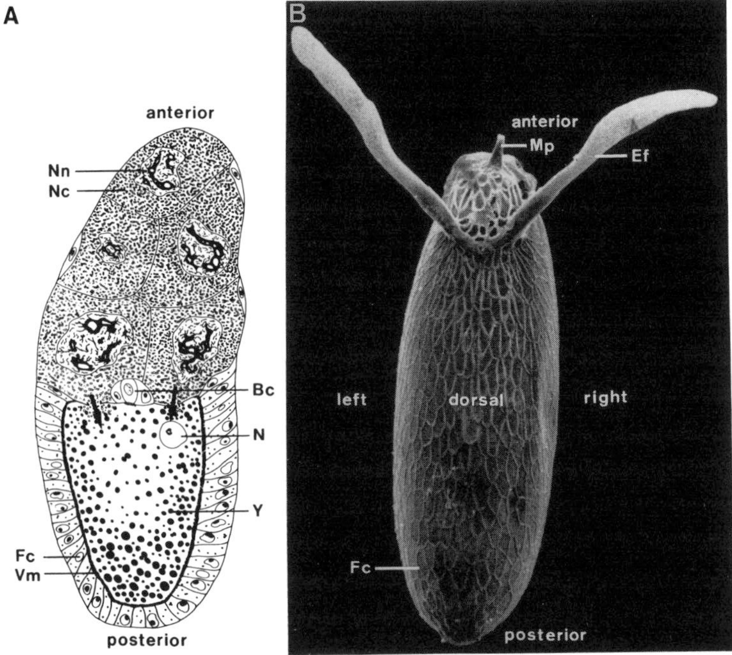

Figure 5.3
Oogenesis in Drosophila. *(A) Ovarian egg follicle surrounded by follicle cells. The contents of the nurse cells are poured through ring canals into the oocyte. From R. King,* Ovarian Development in Drosophila melanogaster *(New York: Academic Press, 1970). (B) Mature egg. Traces of the follicle cells are visible on the outer egg shell.* Abbreviations: Bc *border cells,* Ef *egg filaments,* Fc *follicle cell,* Mp *micropyle,* N *nucleus of the oocyte,* Nc *nurse cell cytoplasm,* Nn *nurse cell nucleus,* Vm *vitellin membrane (inner egg shell),* Y *yolk. Courtesy of H. Gutzeit.*

egg follicle, and when the mature egg passes through the oviduct into the uterus, its axes correspond to those of its mother, the anterior end being oriented toward the head of the mother, the dorsal side toward her back, and so on. Egg polarity and specification of the major body axes have therefore been "invented" in the course of fly evolution and passed on from generation to generation for millions of years. Furthermore, this spatial orientation of the egg in the mother ensures fertilization: the fertilized female stores the sperm in a convoluted tubelike organ, the receptaculum seminis, and the opening of the egg shell, the micropyle, through which the sperm has to enter, passes right in front of the receptaculum when moving through the uterus. This ensures an extremely high rate of fertilization.

We do not know how much information is stored in the egg in the form of templates other than DNA, but one needs to remember that the egg cytoplasm is also elaborated under the genetic control of the maternal genome. Oogenesis and early embryogenesis are therefore also amenable to genetic analysis.

How is a *Drosophila* egg made? The first cells to be segregated in early *Drosophila* development are the pole cells, the future germ cells (see fig. 4.10). As development proceeds, the pole cells migrate into the interior of the blastoderm and colonize the future gonads, either testes or ovaries, where they become stem cells. Stem cells have the capacity to divide asymmetrically into a daughter cell that differentiates and a daughter cell that remains a stem cell. In the adult ovary these stem cells are located at the tips of the ovarioles, where they continue to produce a string of oogonia that develop into mature eggs in assembly-line fashion. The oogonium undergoes four mitotic divisions, leading to sixteen cells (one oocyte, or egg, plus fifteen nurse cells). This group of cells is surrounded by follicle cells of mesodermal origin and forms an egg follicle. The oocyte is always localized posteriorly in the follicle. The nurse cells are connected to each other and to the oocyte via cytoplasmic channels, so that they can actively participate in the development of the egg by feeding cytoplasm through the channels into the egg (fig. 5.3A). Macromolecules and even such organelles as ribosomes and mitochondria required for protein synthesis and respiration can pass across the channels. At the end of oogenesis the nurse cells empty their contents into the egg and the follicle cells secrete the egg shells and die. Their hexagonal outlines remain visible on the outer egg shell (fig. 5.3B). The axes of the future embryo are clearly reflected by the shape of the egg shell, the micropyle marking the anterior pole, the two filaments the dorsal and the left-right sides. Polarization is also visible, however, in the interior in the egg cytoplasm, for example, by the presence of the posterior polar cytoplasm with its characteristic polar granules.

Classical embryological experiments involving ligation, puncturing, localized destruction, or removal of egg cytoplasm and centrifugation and transplantation on various larger insects led to the hypothesis that the antero-posterior pattern was set up by two organization centers, one anterior and one posterior. Because these insects are difficult to analyze genetically, however, the nature of these centers remained elusive for more than half a century. The mechanisms of generating positional information along the antero-posterior axis were elucidated largely through the pioneering work of Janni Nüsslein-Volhard. She began with the analysis of the *bicaudal* mutant as a postdoctoral fellow in my laboratory and persevered with this problem down to the molecular level.

The *bicaudal* mutant, first isolated by Alice Bull, has a typical maternal effect: homozygous mutant mothers (carrying two copies of the defective gene) produce defective eggs that develop into lethal larvae in which head and thorax are replaced by a second abdomen in mirror-image symmetry—that is, producing two tails (*bi,* two, and *cauda,* tail) in opposite orientation but no head and no thoracic segments (fig. 5.4). These poor creatures are obviously unsuited for life, but they do reveal important facts about the positional information in the egg. Even though the *bicaudal* embryo is symmetrical with respect to its abdominal segments, pole cells (primordial germ cells) form only at the original posterior pole. This suggested that there are at least three determinants in the egg, an anterior, a posterior, and a germ cell determinant. The first two have now been identified by molecular genetic techniques, but the germ cell determinant (or determinants) remains a challenge for research.

Nüsslein-Volhard and her co-workers demonstrated that removing cytoplasm from the anterior tip of a normal fertilized egg causes a severe reduction in the developing head structures and in some cases their replacement by the posterior-most structures, the telson. Similarly, if cytoplasm from the posterior pole is transplanted into the anterior tip of the egg, head structures are reduced and posterior structures are duplicated. A combination of the two procedures yielded the most dramatic effect, generating "*bicaudal*-type" embryos at high frequency. Transplantation of cytoplasm from the central region of the egg had no effect on segmentation. Interestingly, removal of cytoplasm from the posterior pole resulted in the loss of abdominal seg-

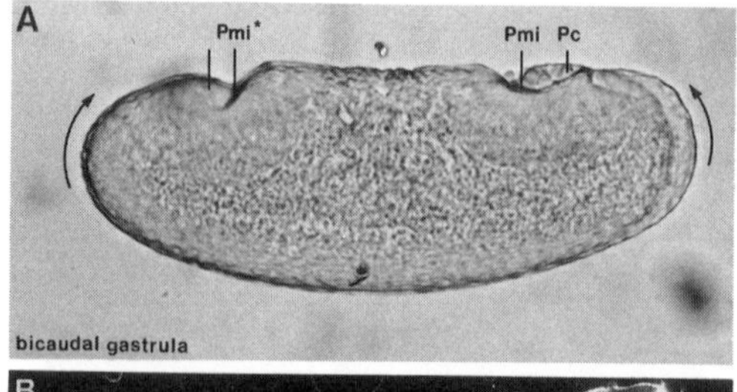

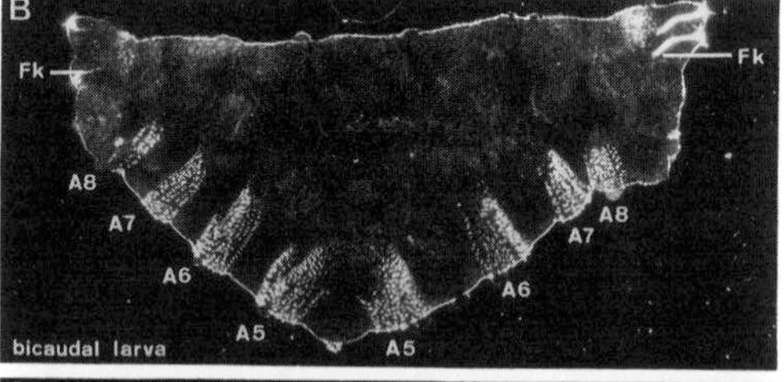

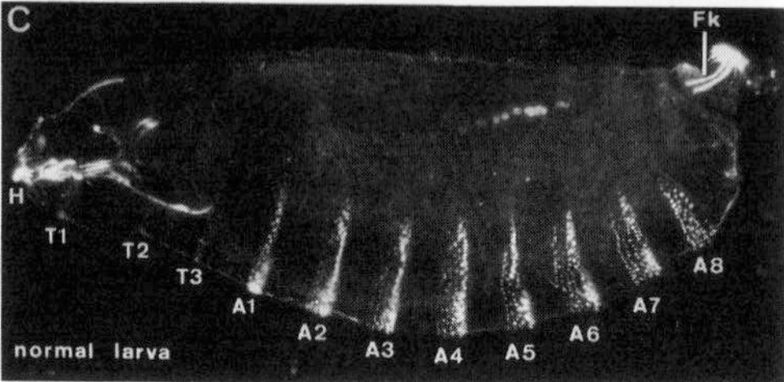

Figure 5.4
bicaudal *mutant. (A)* bicaudal *gastrula. There are two posterior midgut invaginations, a normal one at the posterior pole (Pmi) with pole cells (Pc), and an anterior one (Pmi*) lacking pole cells. (B)* bicaudal *larva with two abdomina in mirror-image symmetry. (C) Normal larva for comparison.* A1–A8 *abdominal segments,* Fk *filzkörper (posterior respiratory openings),* H *head,* T1–T3 *thoracic segments. Courtesy of C. Nüsslein-Volhard.*

ments not immediately adjacent to the posterior pole, whereas removal of cytoplasm destined to become abdomen (at 30 percent egg length from the posterior pole) had little effect on segmentation. These results indicated some long-range effects and suggested a model with two opposing morphogenetic gradients, an anterior one with a maximal morphogen concentration at the anterior pole essential for head and thorax development and a posterior one with its maximum morphogen concentration at the posterior pole required for abdominal development. The most posterior telson (and the most anterior acron) seemed to be controlled differently.

At this point I should mention that this model is not as parsimonious as an engineer or a computer scientist would design it because it implies redundancy. This is nicely illustrated by a discussion I once had with Hans Meinhardt in Tübingen. Meinhardt had made sophisticated computer simulations of the segmentation process based on Alan Turing's reaction-diffusion model, and he told me that in his models he had to assume a posterior

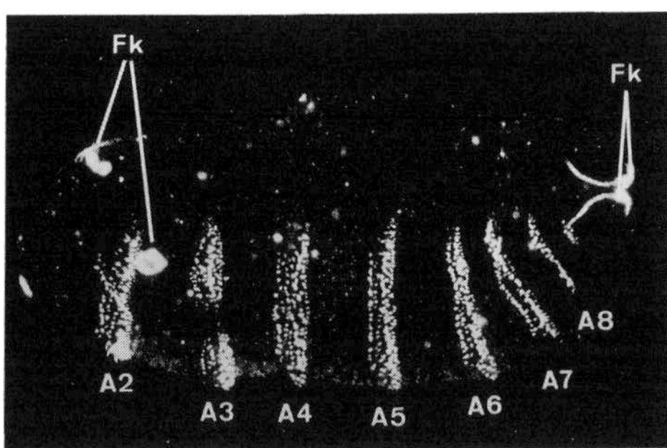

Figure 5.5
bicoid *larva. The head, the thoracic segments (T1–T3), and the first abdominal segment (A1) are missing and replaced by the posteriormost abdominal structures, the filzkörper (Fk). Courtesy of U. Kloter.*

determinant only and that we were "chasing a ghost" by searching for an anterior determinant. This proved to be wrong; Hans Meinhardt was correct in predicting other models on merely theoretical grounds, but in this case nature did not find the most parsimonious solution. As we shall see, evolution involves a lot of tinkering, redundancy, and double assurance that only later is streamlined through selection.

Systematic screenings for maternal effect mutants affecting the anteroposterior segmentation pattern carried out by Nüsslein-Volhard, Trudi Schüpbach, and Eric Wieschaus, who had been my first graduate student, yielded three classes of mutants; the first affecting the anterior system, the second the posterior, and the third the terminal system (both extreme ends, the acron and the telson). The most informative mutant of the anterior system was *bicoid* (*bcd*). Homozygous *bicoid* females produce defective eggs that develop into embryos lacking head and thoracic structures (fig. 5.5). The *bicoid* gene was cloned independently by Gabriela Frigerio in Markus Noll's laboratory and shown to contain a homeobox. Its messenger RNA is synthesized in the nurse cells and transported to the oocyte, where it becomes localized in the anterior end, forming a cap (fig. 5.6). This was the first localized cytoplasmic determinant to be identified. At least three other genes, *exuperantia* (*exu*), *swallow* (*swa*), and *staufen* (*stau*), are required to position the *bicoid* messenger RNA at the anterior pole of the egg. In eggs of *exuperantia* mothers, the *bicoid* messenger RNA is distributed uniformly throughout the egg (fig. 5.6). (Incidentally, the *exuperantia* mutation with its headless phenotype was named after a saint who was beheaded in Zurich in the Middle Ages.) The products of these genes presumably transport and anchor the *bicoid* messenger RNA at the anterior pole. The untranslated trailer segment of the *bicoid* messenger RNA is essential for its localization. Furthermore, the *bicoid* message is masked, or prevented from translation, and becomes unmasked only after fertilization. The detailed mechanisms of localization, masking, and unmasking remain to be worked out.

After fertilization, the *bicoid* message is translated into a homeodomain protein that forms a concentration gradient with its high point at the anterior pole (fig. 5.6). This gradient is abolished in *exuperantia* mutant embryos. The bicoid protein accumulates in the nuclei before the formation of the blastoderm cells. It represents a morphogenetic substance and supplies posi-

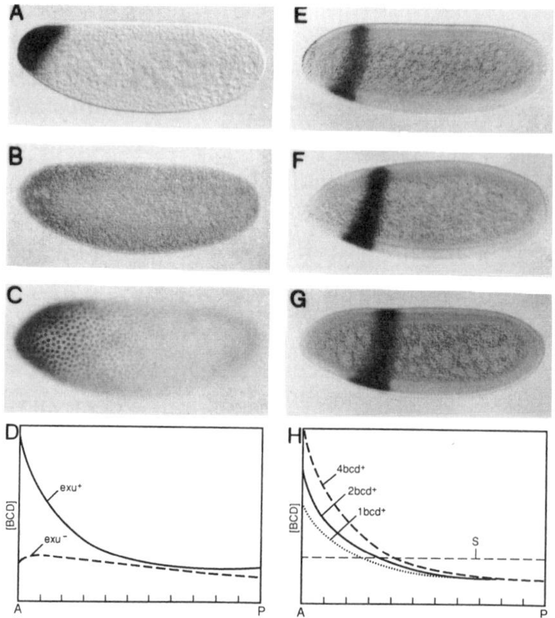

Figure 5.6
Cytoplasmic localization of bicoid (bcd) *messenger RNA and generation of a morphogenetic protein gradient. (A) The maternal mRNA accumulates at the anterior pole of the unfertilized egg. (B) The cytoplasmic localization of mRNA is abolished in* exuperantia *mutant eggs. (C) The translation of the localized* bicoid *mRNA into a protein gradient. The bicoid protein accumulates in the nuclei. (D) The bicoid protein forms a gradient with its highest concentration [BCD] at the anterior pole* (exu$^+$)*. In* exuperantia *mutant embryos* (exu) *the gradient is abolished. (E–G) Expression of the target gene* empty spiracles *in a single anterior stripe. The stripe is progressively shifted in the posterior direction with increasing doses of the* bcd$^+$ *gene (1, 2, and 4 copies, respectively). (H) Dependence of the bicoid protein gradient on the* bicoid *gene dosage from 1 to 4 copies. From R. Wehner and W. J. Gehring,* Zoologie, *23d ed. (Stuttgart: Thieme Verlag, 1995), box 3.1.*

tional information to the respective nuclei. The nuclei respond by turning on the respective target genes. One of these target genes is the homeobox-containing gene *empty spiracles* (*ems*). This gene contains in its flanking region a regulatory DNA element to which the bicoid protein can bind and activate the *empty spiracles* gene. In the normal embryo, *empty spiracles* RNA is expressed in a single stripe in the head region of the early embryo (fig. 5.6). *empty spiracles* is not expressed in the absence of bicoid protein. In the presence of increasing doses of *bicoid* genes, from one to four copies, increasing amounts of bicoid protein are synthesized and the stripe of *empty spiracles* is gradually shifted posteriorly. These findings can be interpreted by assuming that a minimum threshold concentration of bicoid protein is required for *empty spiracles* activation (fig. 5.6). In the anterior tip of the egg *empty spiracles* is repressed by the product of the *tailless* (*tll*) gene.

The crucial experiment demonstrating that *bicoid* is indeed the anterior determinant was carried out in Janni Nüsslein-Volhard's laboratory by Wolfgang Driever, who did most of the molecular analysis of *bicoid.* Driever was able to show that head and thoracic development can be restored in embryos that had some of their anterior cytoplasm removed by the injection of pure *bicoid* messenger RNA. This was a difficult experiment because *bicoid* messenger RNA is rather unstable. However, by replacing the trailer sequences of *bicoid* by those of globin, the message can be stabilized and development of anterior structures restored.

These experiments have resolved the long-standing controversy over whether there are discretely localized cytoplasmic determinants or whether positional information is provided by morphogenetic gradients. In the case of *bicoid,* both views are correct: the RNA is localized discretely at the anterior pole and translated into protein that forms a morphogenetic gradient.

The posterior organization center is considerably more complex. Its molecular genetic analysis began with the isolation and characterization of the homeobox gene *caudal* (*cad*). When carrying out in situ hybridizations of *caudal* DNA to *caudal* messenger RNA in order to localize the transcripts, my former graduate student Marek Mlodzik discovered that the maternally produced *caudal* RNA, which is stockpiled in the egg, is at first uniformly distributed but forms a concentration gradient before blastoderm formation. Embryologists like Sven Hörstadius and Charles Child had long postulated the existence of gradients but had never been able to prove their existence. The *caudal* RNA was the first physical demonstration of such a gradient before the discovery of *bicoid.* Yet the situation in *caudal* is more complicated than that in *bicoid.* The *caudal* RNA gradient is preceded by a stage in which the RNA is uniformly distributed in the oocyte and the newly fertilized egg. At the blastoderm stage, when new transcription sets in, a zygotic transcript is made that differs in structure from the maternal transcript and is expressed as a single belt (or stripe) near the tail end of the embryo. When Mlodzik obtained specific antibodies directed against the caudal protein, he found that the RNA gradient is preceded by a concentration gradient of the caudal protein. The protein accumulates in the nuclei, concentrating at the posterior pole and fading out toward the anterior pole. The existence of the gradient is nicely demonstrated in the *bicaudal* (*bic*) mutant embryo (fig. 5.7). In *bicaudal* mutants, which form two tails with mirror-image symmetry, the gradient is abolished and at the blastoderm stage two symmetrical belts of protein expression appear, one at the original and one at the secondary posterior end. In the oocyte, when the RNA is uni-

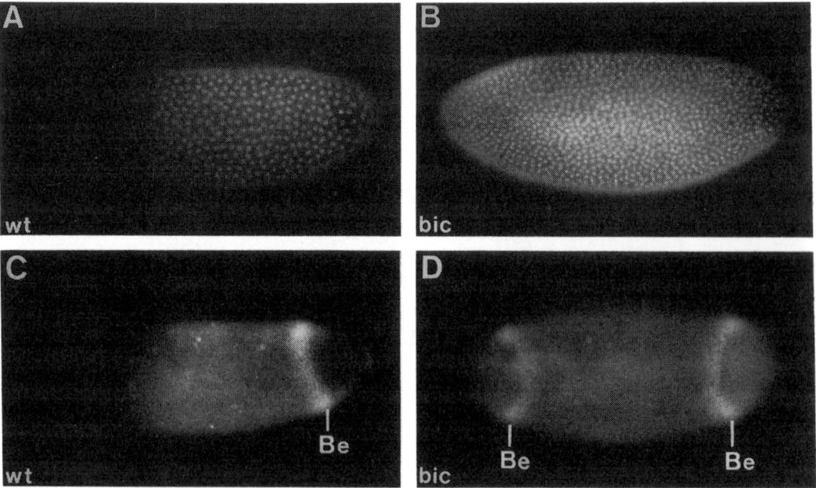

Figure 5.7
The concentration gradient of the caudal protein. (A) caudal protein gradient in a normal (wt) preblastoderm embryo. Note the accumulation of the protein in the nuclei. (B) Uniform distribution in bicaudal (bic) *embryo. (C) caudal protein expression in a posterior belt (Be) in a wild-type (wt) blastoderm embryo. (D) Formation of two belts in mirror-image symmetry in a* bicaudal *embryo. After M. Mlodzik and W. J. Gehring, Expression of the* caudal *gene in the germ line of* Drosophila: *Formation of an RNA and protein gradient during early embryogenesis,* Cell *48 (1987): 465–78.*

formly distributed, the caudal protein is not detectable, suggesting that the RNA is masked. It is not known how the RNA is unmasked and how translation is regulated to form a protein gradient. Mlodzik and I proposed that the bicoid protein, which shows a complementary distribution, might bind to the *caudal* RNA and generate the *caudal* gradient. This hypothesis proved hard to test experimentally, but with the substantial help of Herbert Jäckle and Rolando Rivera-Pomar, we obtained evidence that the bicoid protein does indeed bind to *caudal* RNA. Gary Struhl also obtained evidence independently along the same line. Because mutations in the *bicoid* homeodomain abolish the binding, the homeodomain may mediate not only DNA but also RNA binding. Yet the physical-chemical evidence for RNA binding remains limited. *caudal* mutations isolated by Struhl show surprisingly little effect on abdominal segmentation, with the result that *caudal* has not been considered a major posterior determinant. A *caudal* RNA gradient has also been found in the embryo of the silk moth (*Bombyx mori*), however, suggesting an important function for *caudal* that is conserved through evolution. Because the blastoderm of the silk moth embryo is not syncytial (containing many nuclei in a common cytoplasm), as in *Drosophila,* but subdivided into cells, the formation of a gradient is not precluded by the formation of cell membranes.

A critical gene in specifying the posterior part of the embryo is certainly the maternal-effect gene *nanos* (*nos*). Females homozygous for *nanos* mutations produce embryos that lack abdominal segments but have normal numbers of pole cells. This phenotype confirms the hypothesis that the posterior determinant can be separated from the germ cell determinants. *nanos* encodes a protein with a sequence motif it shares with *vg1* whose transcripts are localized in the vegetal hemisphere of the *Xenopus* egg and are thought to be important for mesoderm induction in the amphibian embryo. The *nanos* RNA is precisely localized in posterior polar cytoplasm for which the products of at least seven other genes of the so-called posterior group are required. Injection of *nanos* RNA into the posterior pole of mutant embryos of any of these posterior group genes rescues their abdominal segmentation defects but not their pole cell defects. The localized RNA serves as a source for the graded distribution of the nanos protein. The major function of this protein is to repress the gene *hunchback* (*hb*) at the posterior pole of the egg. In contrast to *bicoid,* which serves as an activator, *nanos* acts as a repressor, not at the transcriptional but at the translational level. The nanos protein binds to sequences in the trailer region (called nanos-response elements) of the *hunchback* messenger RNA and prevents its translation into protein. The function of *hunchback* is discussed in the following chapter. If these nanos-response elements are tacked onto other messenger RNAs, they also become translationally repressed in the posterior part of the embryo by the nanos protein.

The specification of the anterior and posterior termini is mediated by the maternal-effect genes of the terminal system. The major gene, which for obvious reasons is called *torso* (*tor*), encodes a receptor tyrosinkinase similar to the one described for vulva induction in the worm.

The maternal-effect genes of the anterior, posterior, and terminal systems specify an antero-posterior coordinate system that provides positional information to the nuclei when they migrate from the interior of the egg to the periphery to form the blastoderm.

Also specified by maternal-effect genes are the dorso-ventral coordinates. The dorso-ventral gradient is generated by yet another mechanism that involves nuclear localization. Proteins are synthesized in the cytoplasm on polyribosomes by translation of the respective messenger RNAs. The finished protein molecules are targeted to various sites within the cell, for example, receptors may be targeted to the plasma membrane, whereas transcription factors (activators and repressors) have to be translocated to the nucleus. Nuclear proteins contain specific nuclear targeting sequences required for nuclear translocation. The *dorsal* (*dl*) gene, the key player in specifying dorso-ventral positional information, was again identified in a screen for maternal-effect mutants. Loss-of-function mutants of *dorsal* produce embryos consisting of dorsal epidermis and yolk only. Normal dorsal function is thus required for forming ventral structures. In contrast, *cactus* (*cact*) mutants are ventralized and have an antagonistic effect to *dorsal.* The *dorsal* gene was cloned by my former graduate student Ruth Steward-Silberschmidt and shown to encode a transcription factor. Neither *dorsal* RNA nor dorsal protein are differentially distributed along the dorso-ventral axis of the embryo, but the cell nuclei accumulate dorsal protein to a much higher concentration on the ventral side than laterally, whereas in the most dorsal cells dorsal protein is exclusively located in the cytoplasm. In ventralized embryos, as those produced by *cactus* mothers, dorsal protein accumulates in all nuclei of the blastoderm, whereas in dorsalized embryos it remains exclusively in the cytoplasm. These results are consistent with the idea that *dorsal* (in spite of its name, which refers to the mutant phenotype) is a ventral determinant and that the nuclear gradient of dorsal protein specifies the dorso-ventral coordinate system of positional information. Cactus protein, in contrast, prevents dorsal protein from entering the nuclei and retains it in the cytoplasm, where it is inactive.

In summary, at this early stage of development the egg contains an antero-posterior and a dorso-ventral system of morphogen gradients that establish a coordinate system of positional information that is communicated to the nuclei during blastoderm formation as a first step in establishing the body plan.

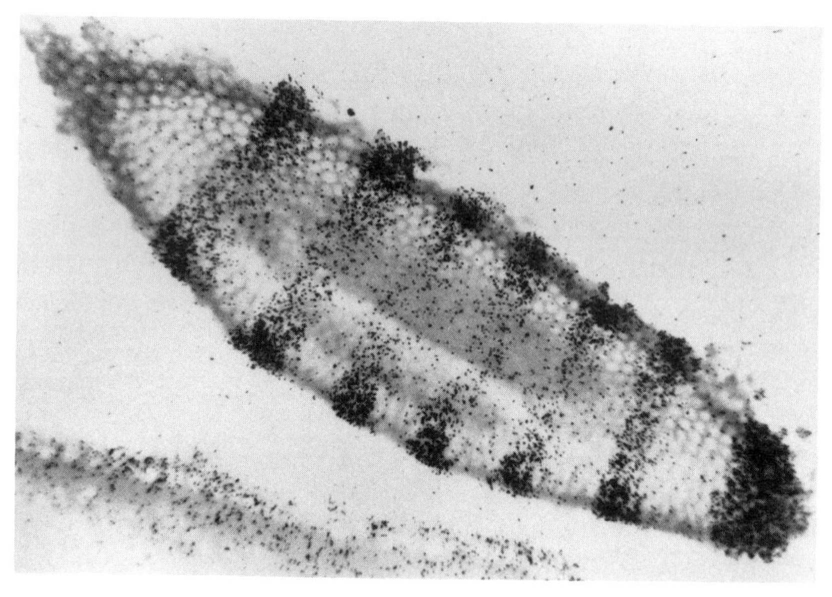

Figure 6.1
The seven stripes of expression of the segmentation gene fushi tarazu. *Courtesy of E. Hafen.*

6
From Gradients to Stripes

Ernst Hafen and Atsushi Kuroiwa stormed into my office and dragged me to the microscope in the laboratory. They had just developed the first in situ hybridization slides using *fushi tarazu* DNA as a probe. The result was spectacular: *fushi tarazu* was expressed in a beautiful pattern of seven stripes, or belts, around the embryo (fig. 6.1). The gene was first identified as a mutation by Barbara Wakimoto, a second-generation Japanese-American, who gave it a Japanese name that means not enough segments. *fushi tarazu* mutant embryos lack every other segment (or more precisely every other parasegment) and are therefore lethal (fig. 6.2). The mutation maps close to *Antennapedia*, and Kuroiwa cloned the gene on the basis of its homeobox.

Around the same time, Hafen and Michael Levine had improved the method of in situ hybridization so that it could detect homeotic gene transcripts, and they applied it first to *Antennapedia.* When Kuroiwa had prepared suitable *fushi tarazu* probes, he asked me about doing in situ hybridization experiments in collaboration with Hafen and predicted that this gene might be expressed in stripes, assuming that it is expressed in those areas of the wild-type embryo that are missing in the *fushi tarazu* mutant. I

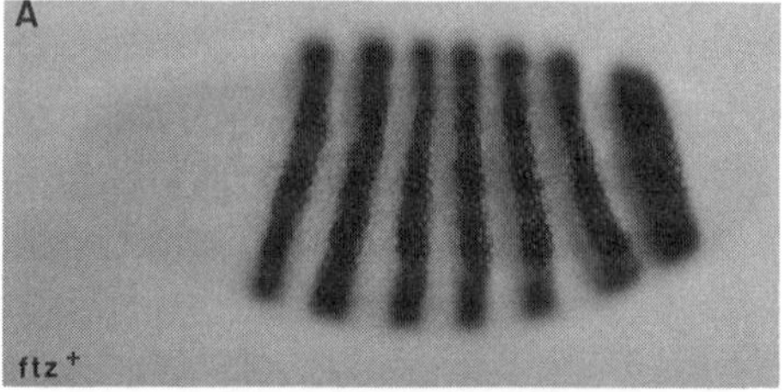

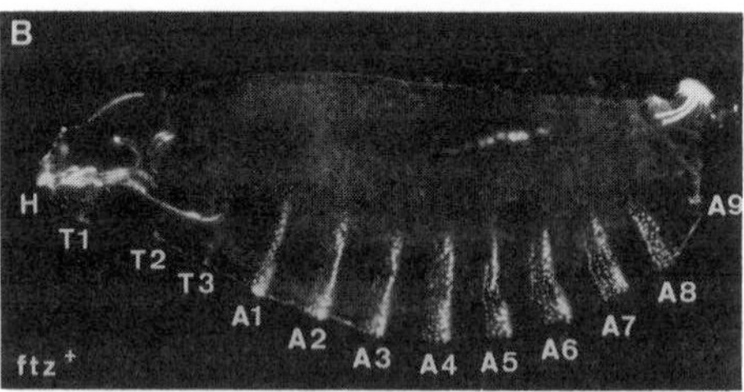

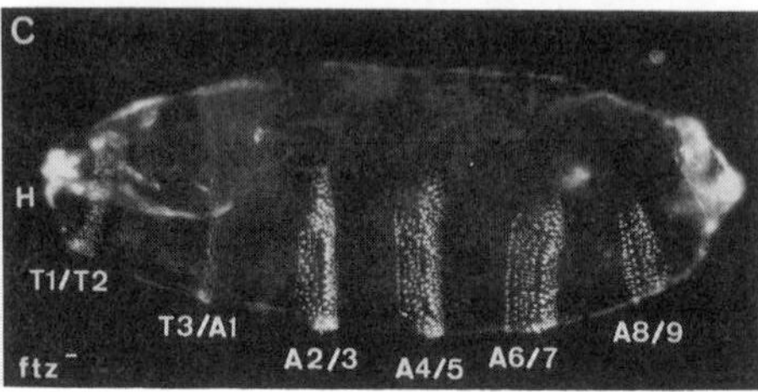

Figure 6.2
fushi tarazu *gene expression and mutant phenotype: (A) expression of* fushi tarazu *messenger RNA in seven stripes in a normal* (ftz$^+$) *embryo; (B) segmentation pattern of a normal first-stage larva; (C) segmentation pattern of a* fushi tarazu (ftz$^-$) *mutant larva. Alternate parasegments are missing, and the remaining parts of the body segments (e.g., anterior T3 and posterior A1) are fused.*

was excited about this idea, but I warned him not to be disappointed when the expression pattern would differ from his expectations, because the RNA transcripts would be uniformly distributed if gene regulation occurred at the RNA or protein rather than the DNA level. He remained optimistic, however, and joined forces with Hafen to carry out the proposed experiment. To our great excitement, his prediction was correct. Here they were, the stripes! It was one of those rare moments that are highlights in the life of a scientist. The stripes exactly reflected the map of the embryo as derived from laser ablation experiments. The fate map was "painted" onto the blastoderm. Even though all the blastoderm cells, with the exception of pole cells, look absolutely alike, they differentially express the *fushi tarazu* gene. Even before blastoderm formation, when the nuclei become separated by cell membranes, they differ with respect to gene expression; the transcripts are localized in stripes, indicating that the RNA cannot freely diffuse and that more than the cell membrane prevents them from diffusion. This came as a surprise to some of my biochemist friends, who convert an embryo into a soluble extract before studying it. Furthermore, the fact that the pattern formed before cellularization indicated that it was not generated by cellular interactions and implied that it was induced by determinants or morphogenetic substances located in the cortical (peripheral) egg cytoplasm that presumably enter the immigrating nuclei and activate the *fushi tarazu* gene locally.

Finding the stripes was a breakthrough in the analysis of segmentation gene action, and I immediately telephoned Janni Nüsslein-Volhard to tell her the news. Unfortunately, on that day Janni was not in Tübingen, and Gerd Jürgens, one of her collaborators, answered the phone. He was rather skeptical and did not share my excitement, but he promised at least to tell Janni when she returned. She and Eric Wieschaus had embarked upon a comprehensive screen for embryonic lethal mutations affecting segmentation. Their large collection of mutants could be subdivided into various classes that turned out to be essential for understanding the genetic control of segmentation. The first class, called gap mutants, affects several adjacent body segments; the second, the pair-rule mutants, affects every other segment or parasegment (with double-segment periodicity); and the third class, designated as segment polarity mutants, leads to alterations in every segment. *fushi tarazu* belongs to the class of pair-rule mutants, and Janni kindly gave us several *fushi tarazu* mutant stocks without which the further analysis may well have been impossible.

The first in situ hybridizations were carried out on sections through embryos and the three-dimensional pattern of RNA expression had to be reconstructed. Fortunately, the pattern of RNA stripes was immediately obvious in tangential sections (see fig. 6.1). But we wanted to find out whether indeed the mutant defects coincided precisely with the pattern of RNA expression in the wild type. Examination of the *fushi tarazu* mutant phenotype revealed that it was not every other segment that was missing but rather the posterior portion of one segment and anterior portion of the next segment—that is, a unit defined as a parasegment by Peter Lawrence and Alfonso Martinez-Arias. Our in situ hybridization data clearly indicated that *fushi tarazu* is expressed in parasegments rather than segments. Methods for in situ hybridization to whole embryos with nonradioactive probes were later developed that greatly facilitated this analysis (see fig. 6.2). In particular, double-labeling with *engrailed* (*en*), a gene expressed in the posterior compartment of each segment (see figs. 6.10, 6.11) confirmed the parasegmental nature of *fushi tarazu* expression.

To study protein expression, Henri Krause, a postdoctoral fellow in the laboratory, prepared antibodies against fushi tarazu protein synthesized in bacteria. The pattern of protein expression essentially coincided with that of

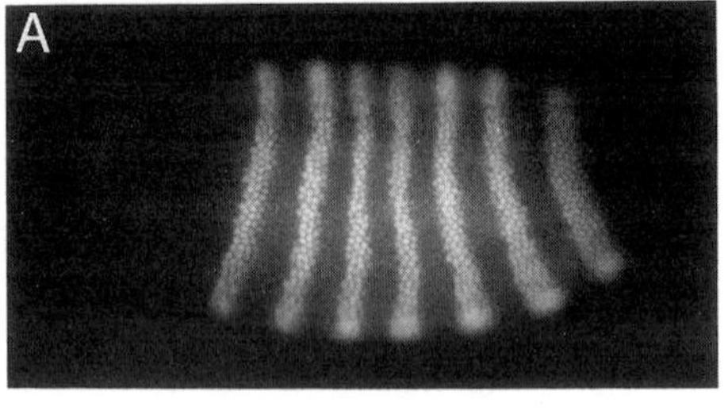

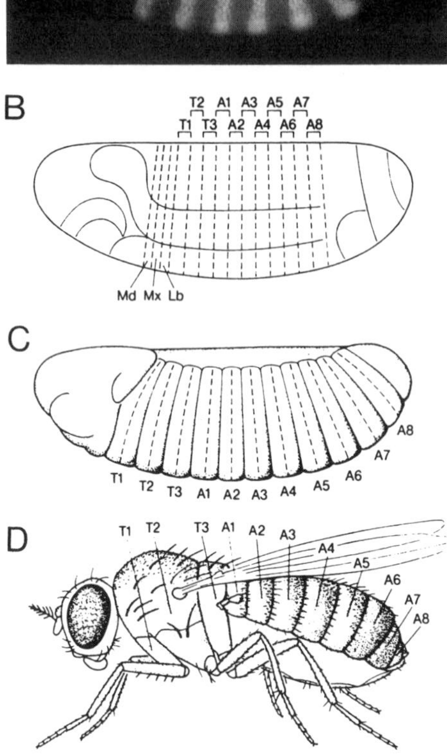

Figure 6.3
The body plan is "printed" on the blastoderm embryo as reflected by fushi tarazu *expression: (A) expression pattern of fushi tarazu protein as revealed by antibody staining; (B) fate map of the blastoderm embryo; (C) morphological appearance of the body segments at a later embryonic stage; (D) segmentation pattern of the adult fly.*

RNA, suggesting that control was exerted largely at the transcriptional level. Sean Carroll and Matt Scott obtained similar results. As expected for a transcription factor, fushi tarazu protein is localized primarily in the nuclei (fig. 6.3). Another dogma was about to be shattered, one concerning the number of gene regulatory molecules per cell. Studies of the bacterial lac repressor indicated that each cell contains very few repressor molecules, just enough to interact with that single gene that must be regulated. One skeptic in this respect was Don Brown, who asked me at a Cold Spring Harbor meeting how many fushi tarazu protein molecules there were per cell. When I told him that we estimated between ten thousand and twenty thousand, he remarked that fushi tarazu was unlikely to be a gene regulatory protein.

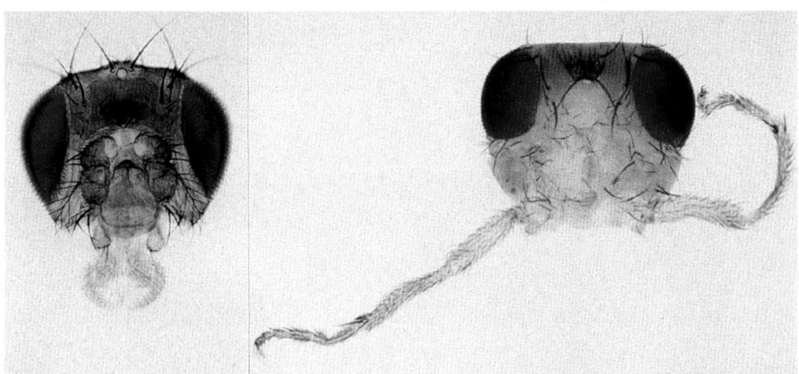

Plate 1

Antennapedia. *The head of a normal (wild-type) fruit fly on the left is compared to the homeotic* Antennapedia *mutant head on the right. In the mutant the antennae are transformed into middle legs.*

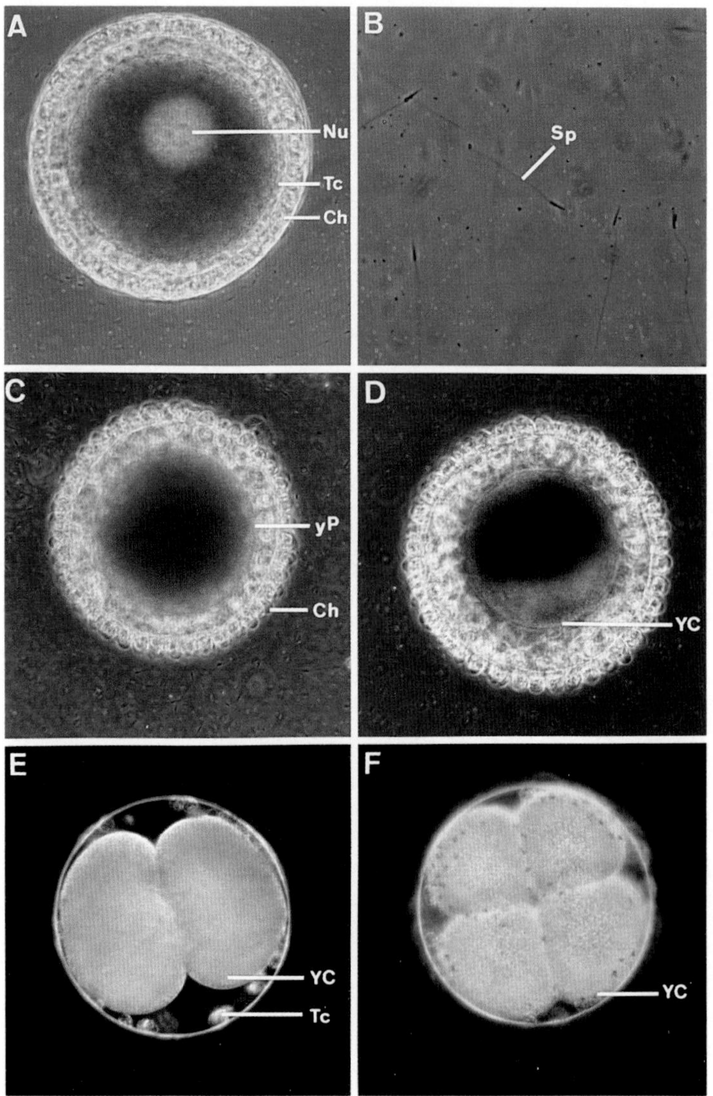

Plate 2

The yellow crescent in the egg of the sea squirt Styela plicata. *(A) Oocyte with nucleus (Nu), surrounded by test cells (Tc) and chorion cells (Ch); (B) Sperm cells; (C) Unfertilized egg with an even distribution of yellow pigment granules (yP) around the periphery of the egg; (D) Fertilized egg. Formation of the yellow crescent (YC) upon fertilization; (E) Two-cell stage. Bisection of the yellow crescent by the first cleavage division; (F) Four-cell stage. Segregation of the yellow crescent material into two of the four cells.*

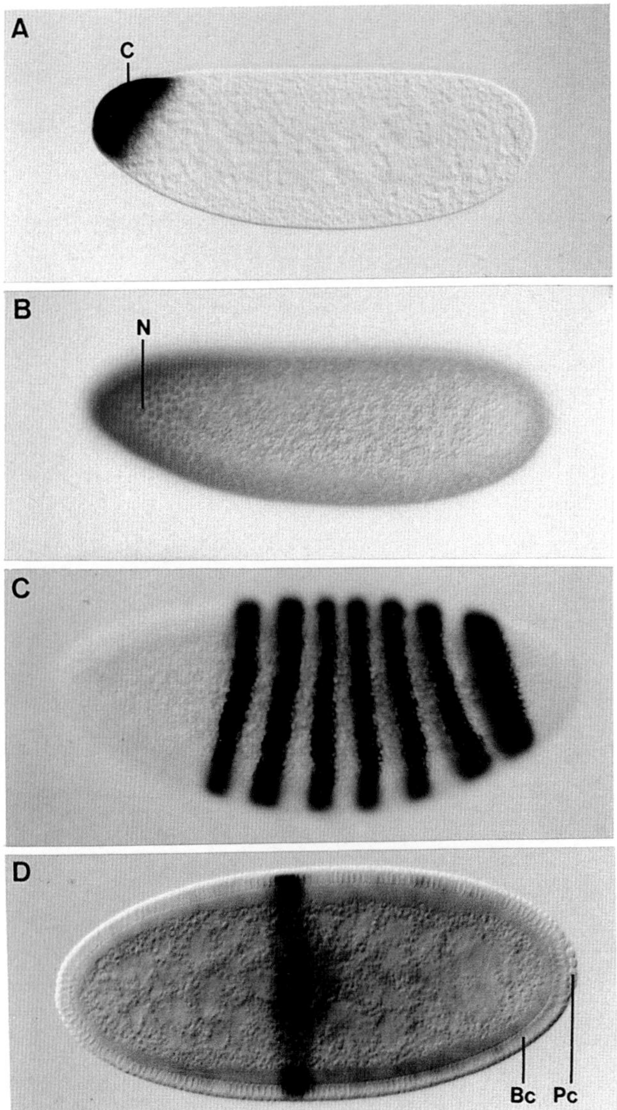

Plate 3

Four conceptual steps in pattern formation during embryonic development of Drosophila: *(A) cytoplasmic localization of* bicoid *messenger RNA at the anterior pole of the egg in a cap (C); (B) formation of a morphogenetic gradient of bicoid protein with its highest concentration in the nuclei (N) at the anterior pole of the preblastoderm embryo; (C) expression of the* fushi tarazu *messenger RNA in a repetitive pattern of seven stripes at the preblastoderm stage; (D) expression of the* Antennapedia *mRNA in a unique pattern of a single stripe at the blastoderm stage.* Bc *blastoderm cells,* Pc *pole cells. Anterior is always to the left and dorsal at the top* (A, *C, and D in situ hybridizations, B antibody staining).*

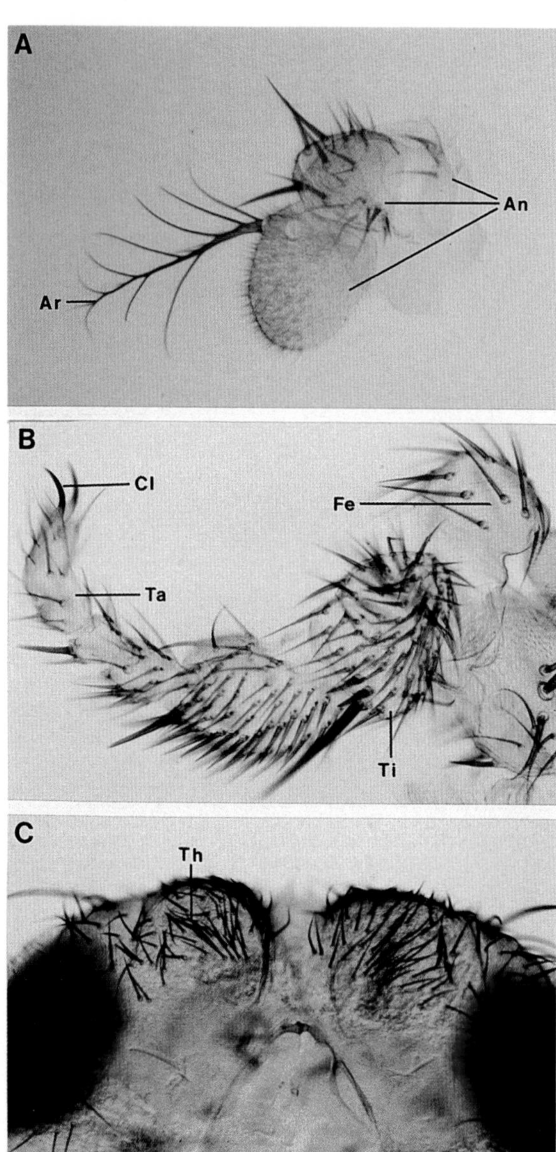

Plate 4

Homeotic transformation of the antennae into middle legs and the dorsal head region into mesothoracic structures by ubiquitous expression of Antennapedia: *(A) normal antenna,* An *three antennal segments,* Ar *arista; (B) antenna-to-leg transformation,* Cl *claws,* Ta *tarsus,* Ti *tibia,* Fe *femur; (C) head-to-mesothorax transformation,* Th *thoracic bristles.*

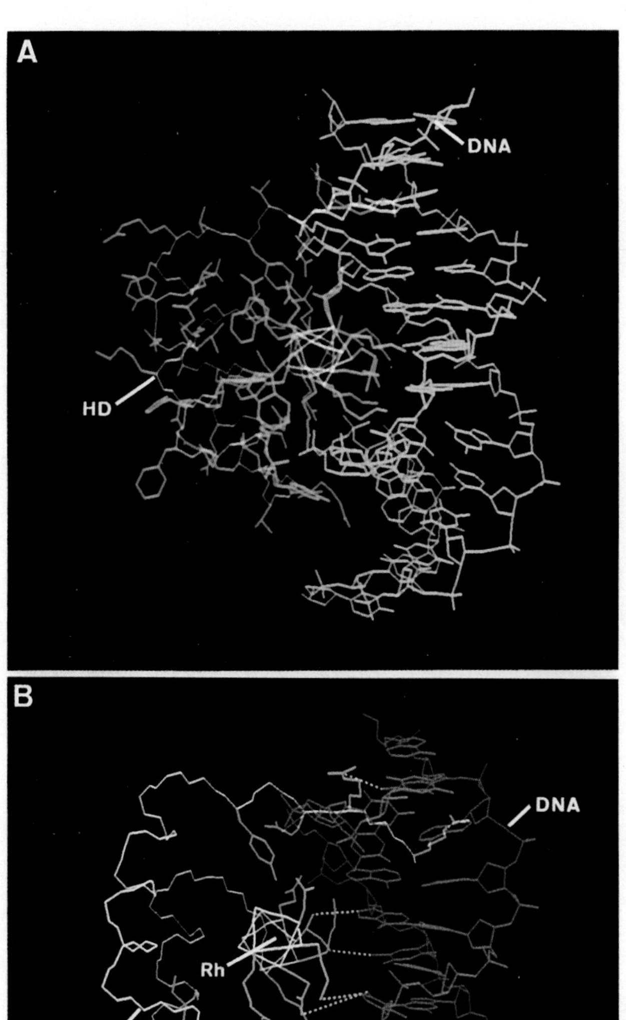

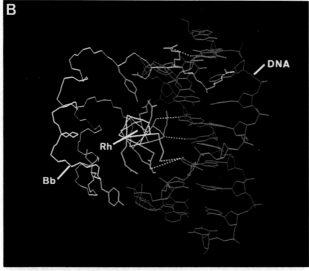

Plate 5
Structure of the Antennapedia *homeodomain-DNA complex as determined by nuclear magnetic resonance spectroscopy. (A) Structure with all side chains represented. DNA (yellow), homeodomain (HD), backbone (magenta), amino acid side chains (blue). (B) Homeodomain-DNA contacts. DNA (red), homeodomain backbone (Bb, magenta), amino acid side chains contacting DNA (blue), direct contacts between amino acids and specific bases (yellow dotted lines). Courtesy of K. Wüthrich and M. Billeter.*

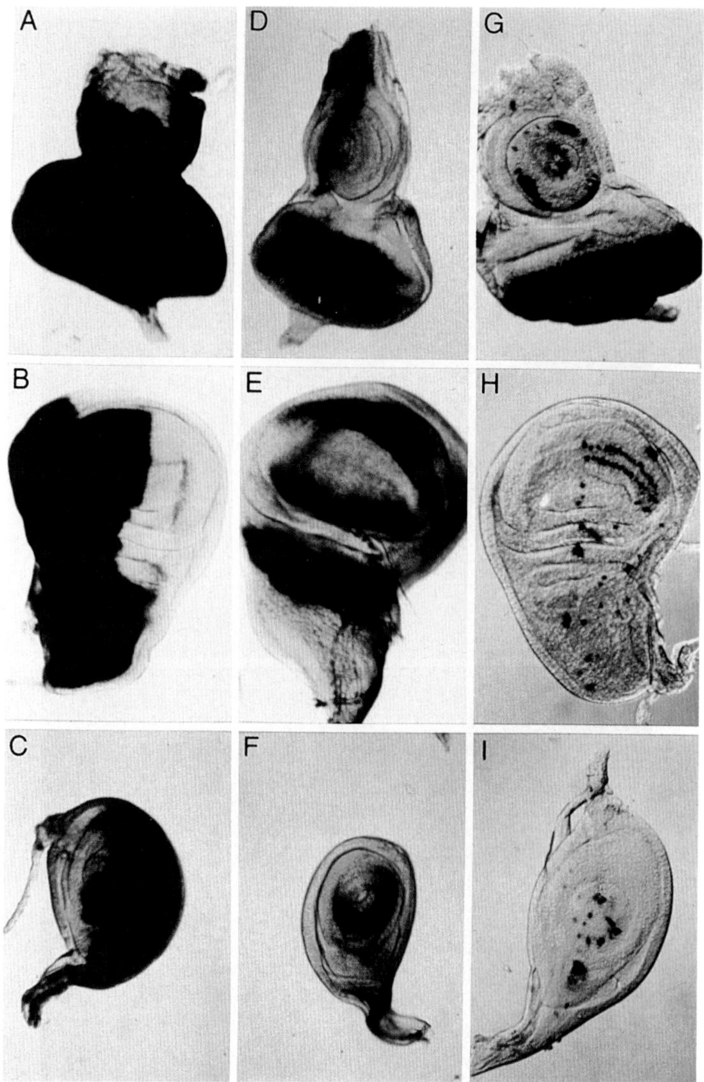

Plate 6

Expression patterns of various enhancer detector lines. A, D, G *eye-antennal discs;* B, E, H *wing discs;* C, F, I *leg discs.* A–C *enhancer detector line expressed specifically in the anterior compartment of the imaginal disc;* D–F *horseshoelike pattern of expression in the antenna, wing, and leg discs;* G–I *enhancer detector line expressed specifically in sensory organs in the antenna, wing, and leg discs and in the photoreceptor cells of the eye disc. The blue stain reflects the expression of the* β*-galactosidase reporter gene under the influence of resident enhancers. Courtesy of G. Gibson.*

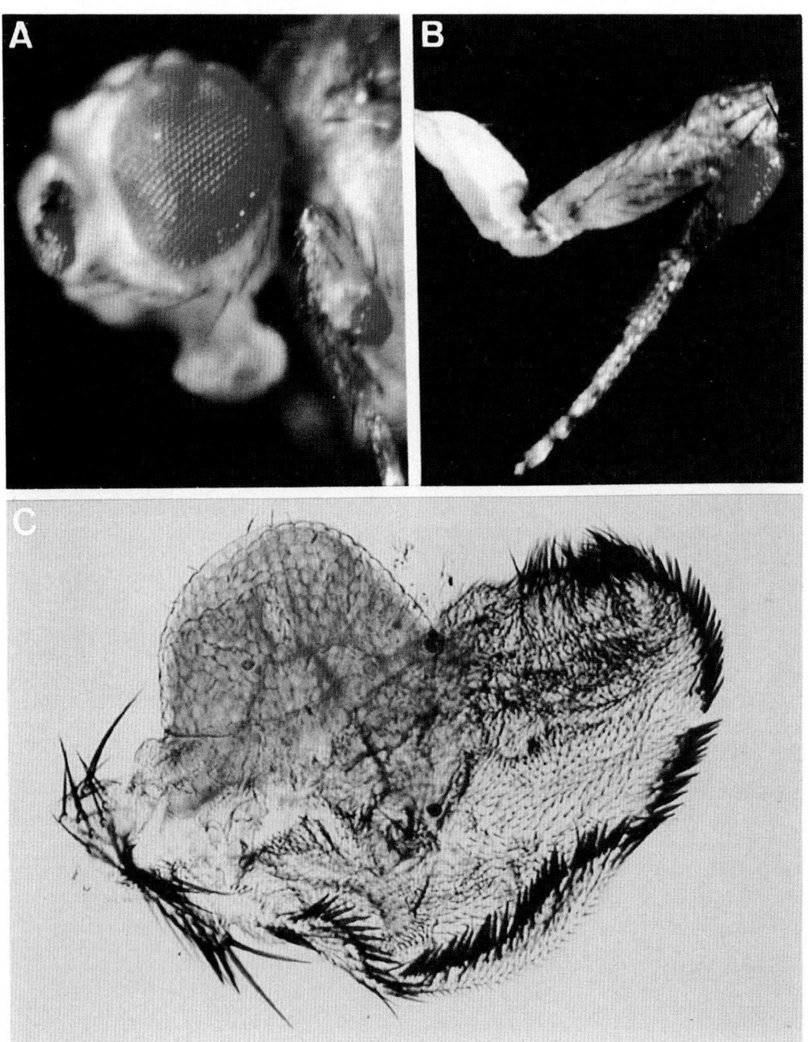

Plate 7

Induction of additional eyes on the antennae, legs, and wings of a transgenic fly by expression of the normal eyeless *gene: (A) additional eyes on the antenna and the foreleg; (B) higher magnification of an eye induced on the tibia of the foreleg; (C) large eye on wing. From G. Halder, P. Callaerts, and W. J. Gehring, Induction of ectopic eyes by targeted expression of the* eyeless *gene in* Drosophila, Science *267 (1995): 1788–92, with permission from the American Association for the Advancement of Science (AAAS), Washington, D.C.*

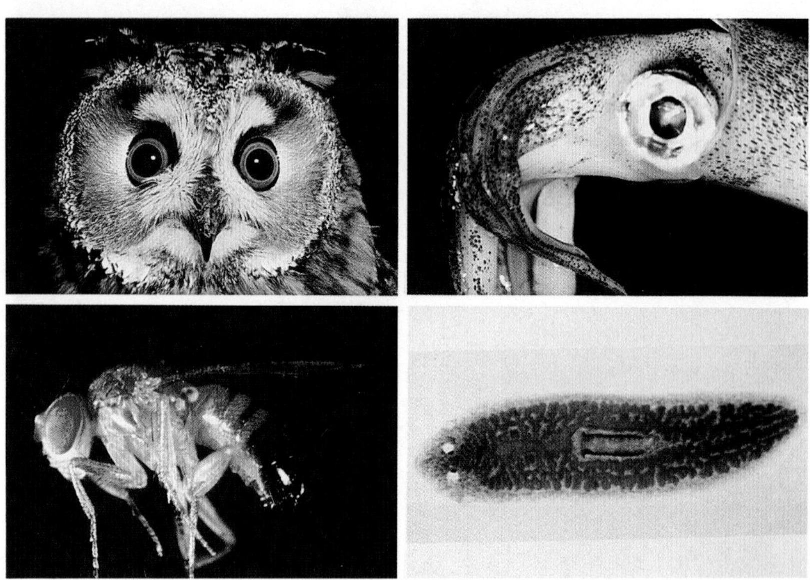

Plate 8
Various eye types in animal phyla; top row: *vertebrates (owl) and cephalopods (squid);* bottom row: *insects (fruit fly) and flatworms (Planaria). Owl courtesy of E. Zbären; squid and fruit fly photographs by the author.*

Higher organisms have much larger genomes than bacteria, however, and they solve the problem of finding a single gene in a huge genome by increasing the number of regulatory protein molecules rather than by increasing their affinity for the DNA binding site beyond that which is found in bacterial repressors.

Immunolocalization experiments showed that after being expressed in stripes, the fushi tarazu protein was expressed in the ventral nervous system, but in this case in every segment and not in every other as for the stripes. The pattern of expression in specific nerve cells is extremely precise and bilaterally symmetrical, indicating that gene expression is regulated precisely at the single-cell level and is not fuzzy. Later in development, *fushi tarazu* is expressed in a ring of gut cells, but the expression of the gene is confined to the embryonic stage.

The best way to prove that you have isolated the right gene that corresponds to the respective mutation is to introduce the cloned DNA into the germline of transgenic flies and to test whether it complements or rescues the mutation. Yash Hiromi, following up on Kuroiwa's work, set out to do that experiment. *fushi tarazu* appeared to be a small gene as compared to *Antennapedia*, which spans more than 100,000 base pairs and is much harder to handle. The *fushi tarazu* coding region measures about 2 kb only (1 kb = 1,000 base pairs) and consists of two exons separated by a small intron. As usual, the homeobox is located in the second exon, encoding the C-terminus of the protein, close to the intron-exon junction. Hiromi inserted this piece of DNA into a P-element vector, transformed flies with it, and crossed them back into a *fushi tarazu* mutant background, but without success; the cloned piece of DNA did not rescue the defect caused by the mutation. I encouraged him to insert progressively longer and longer pieces of DNA, but the exercise was unsuccessful until he added 6 kb of DNA upstream to the coding region. At this point, the mutation was rescued, and he obtained viable embryos with the normal number of segments—a nice case of gene therapy (fig. 6.4). Obviously, the control region of this gene is highly important and comprises three times more DNA than the coding region. In some genes the regulatory elements are also found in the introns or at the 3' end, but in *fushi tarazu* the regulatory elements were confined to the 5' side upstream of the gene, except for a more quantitative element located downstream.

These regulatory elements were subsequently analyzed by fusing the 6 kb upstream region to a reporter gene, encoding bacterial β-galactosidase, a method originally developed by John Lis. In transgenic embryos carrying

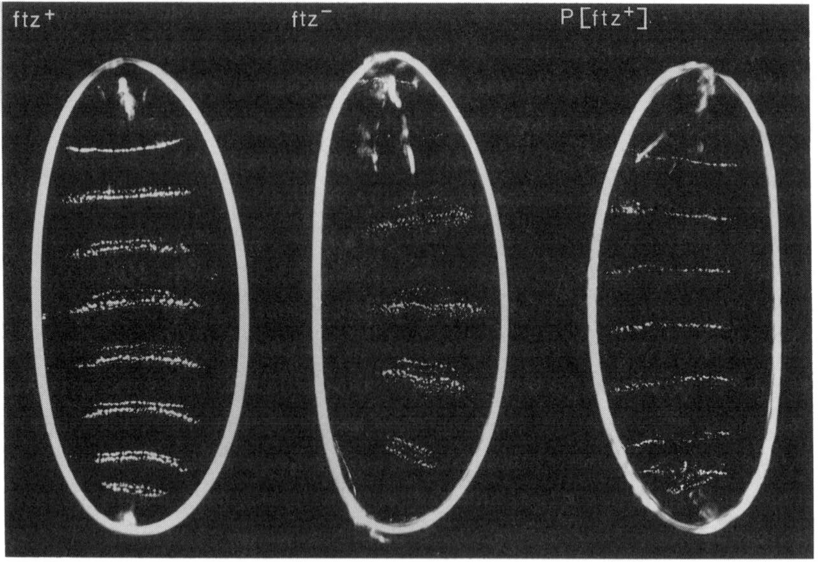

Figure 6.4
Rescue of the fushi tarazu *mutant by transfer of the wild-type* (ftz^{+}) *gene. A normal* (ftz^{+}) *embryo, a mutant* (ftz) *and a rescued (P[*ftz^{+}*]) embryo (surrounded by the vitellin membranes) are shown. The* ftz^{+} *gene was inserted into a P-transposon vector and injected into eggs. The P[*ftz^{+}*] transgene is capable of restoring the normal number of segments in a* ftz^{-} *mutant.*

such fusion constructs, gene activity can be easily detected by assaying for β-galactosidase activity. For this purpose a substrate is provided that is enzymatically converted by β-galactosidase into a blue-colored precipitate. These experiments, which were initiated by Kuroiwa and carried out by Hiromi with great success, identified four well-defined regulatory elements upstream of the start site: the "zebra" element, responsible for the formation of the seven stripes, the "neurogenic" element, controlling the expression in the nervous system, a dispensable region, and finally an enhancer responsible for the enhancement of the stripes (fig. 6.5).

When fused to the reporter gene, the zebra element and the enhancer are capable of directing the expression of β-galactosidase in seven stripes, the first β-galactosidase "zebra" ever generated, whereas the neurogenic element directs expression of β-galactosidase in the precise pattern of the ventral nervous system. The sequence of these upstream regulatory elements was determined for different *Drosophila* species in collaboration with Dieter Maier, and it turned out that the regulatory sequences, with the exception of the dispensable region, were highly conserved in evolution even more strongly than parts of the protein coding region. This indicates that there is a strong selective pressure on these regulatory DNA sequences so that mutations are generally not tolerated.

I describe these experiments in some detail so that the reader can appreciate the principles of gene regulation. Not only does a gene have to be turned on or off, but it has to be fine-tuned and expressed in a precise spatial and temporal pattern in a concerted action with thousands of other genes. It is a musician in a large orchestra that is playing the symphony of life. Directing the orchestra are master control genes, and the full score, the program, is written into the DNA.

Most genes are endowed with two basic types of regulatory elements, promoters and enhancers. The promoter is located immediately upstream of the start site of transcription. It contains a consensus DNA sequence TATAAA called the Hogness-Goldberg box (or TATA box), first identified by David Hogness. In 1978–79 Hogness spent a sabbatical year in my laboratory at the Biozentrum in Basel and discovered the TATA box while analyzing the DNA sequences of the *Drosophila* histone genes determined by his graduate student Michael Goldberg. We have subsequently baptized the room in which the discovery was made as the "Hogness Box." The TATAAA

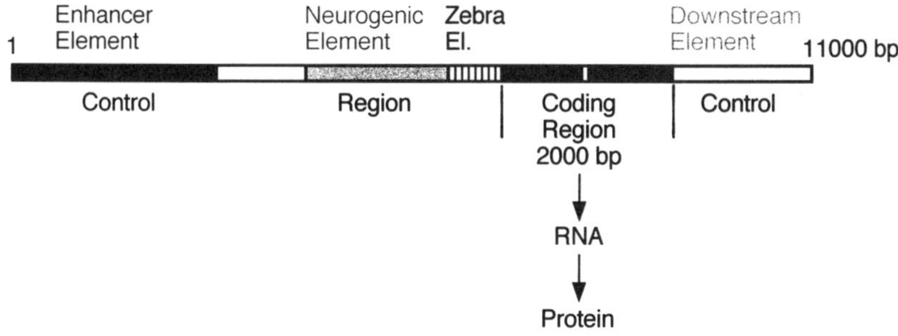

Figure 6.5
Structure of the fushi tarazu *gene. The upstream control region consists of an autoregulatory enhancer element, a neurogenic element, and the zebra element, which also comprises the promoter. It is followed by the coding region and the downstream element.*

sequence is recognized by a DNA binding protein, the TATA binding factor, which positions the RNA polymerase complex on the start site in order to initiate transcription—that is, the synthesis of the messenger RNA. The enhancers can be located upstream, inside, or downstream of the coding region of the gene and also represent binding sites for proteins that regulate transcription. It is generally assumed that proteins bound to the enhancers simultaneously bind to proteins associated with the promoter and that the DNA in between forms a loop.

The genes acting upstream of *fushi tarazu* in the hierarchy of the control genes can be identified by looking at *fushi tarazu* expression in the respective mutant embryos or by crossing these reporter gene constructs into a mutant

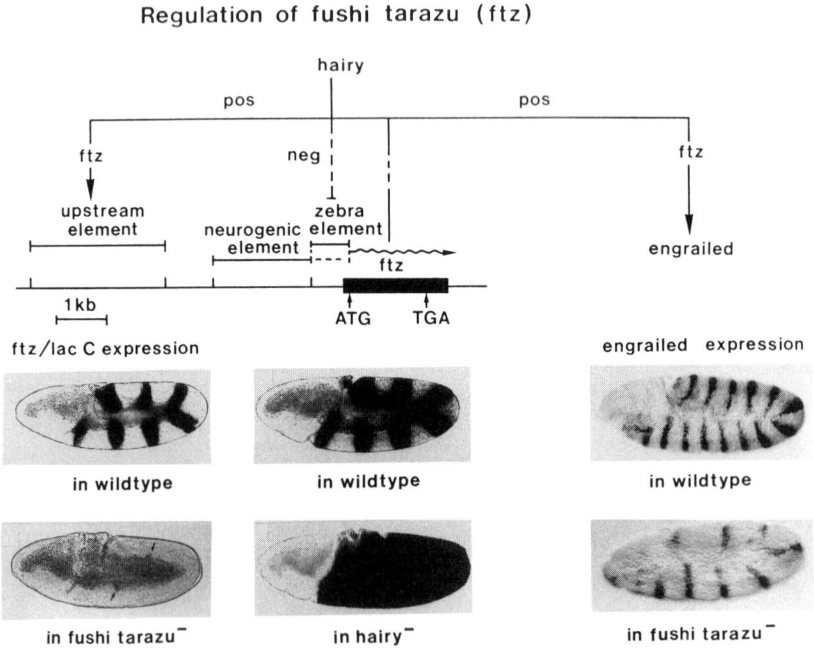

Figure 6.6
fushi tarazu *regulation. The* ftz *coding region is shown as a black bar, the regulatory regions (upstream element, neurogenic element, and zebra element) as horizontal lines. The wavy line indicates the* ftz *transcript.* ftz *activates (pos)* engrailed *and positively autoregulates itself.* ftz *expression is repressed (neg) by* hairy, *which interacts with the zebra element.* Lower part: *Expression of* ftz/lacC *fusion gene in which the coding region of* ftz *is replaced by the β-galactosidase reporter gene. The reporter gene is expressed in seven stripes in the wild type, whereas in* ftz⁻ *embryos the stripes are missing. In* hairy⁻ *embryos, the reporter gene is also expressed in interstripes. In wild-type embryos* engrailed *is expressed in 14 stripes, whereas in* ftz⁻ *embryos alternate stripes are missing.*

background. If, for example, the zebra element construct is transferred into a *hairy* (*h*) mutant embryo, the seven stripes become much broader and the regions between the stripes also express *fushi tarazu* (fig. 6.6), so that only the head region remains unstained. This suggests that *hairy* normally represses *fushi tarazu* in the inter-stripes; if *hairy* is inactive, *fushi tarazu* becomes derepressed. This interpretation is consistent with the finding that *hairy* is also expressed in seven stripes, but the *hairy* stripes are located in the inter-stripes of *fushi tarazu.* Apparently the preblastoderm nuclei are sending signals to each other, one group producing *fushi tarazu,* the adjacent group expressing *hairy.*

Because there are no barriers between the nuclei, transcription factors can directly be used for signaling. Like *fushi tarazu, hairy* also encodes a transcription factor not with a homeodomain but rather with another DNA binding domain, a helix-loop-helix motif. However, these experiments do not prove a direct interaction with the hairy protein binding to the *fushi tarazu* regulatory element. The action of *hairy* may also be indirect, via another transcription factor. A molecular genetic method for demonstrating a direct interaction is explained later.

Long-range effects of *bicoid* on the fate map can also be demonstrated by crossing the *fushi tarazu*-β-galactosidase fusion construct into *bicoid* mutant embryos. The entire fate map is shifted anteriorly, and some of the anterior *fushi tarazu* stripes are fused or missing (fig. 6.7).

When the enhancer fusion construct, which in the normal (wild-type) embryo directs expression in the seven stripes, is crossed into *fushi tarazu* mutant embryos, the stripes do not appear and there is practically no expression at all (see fig. 6.6). This suggests an autoregulatory (self-regulatory) function: normally the fushi tarazu protein binds to its own enhancer and enhances its own expression in the seven stripes, amplifying its expression or that of its reporter gene. This autoregulatory feedback loop is interrupted in *fushi tarazu* mutants that fail to produce active protein. This model of autoregulatory feedback was later proven by Alexander Schier, who demonstrated the direct binding of fushi tarazu protein to its enhancer in vivo. Such autoregulatory feedback systems were predicted by Hans Meinhardt on theoretical grounds.

The enhancer also contains multiple binding sites for several transcription factors other than for the fushi tarazu protein. These transcription factors are not *fushi tarazu* specific; they also control sets of other genes. This leads to a combinatorial model for gene regulation, which resembles a tele-

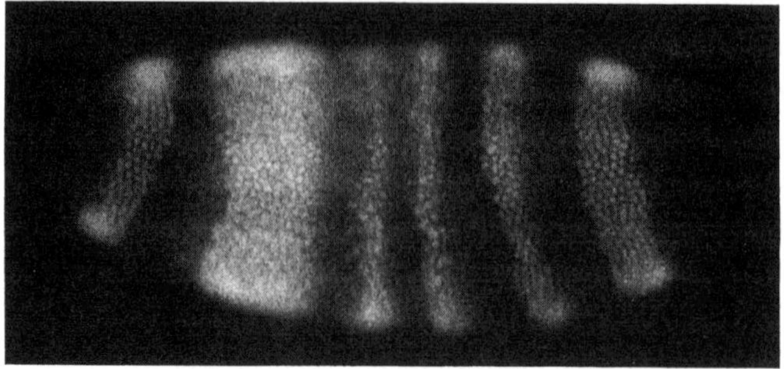

Figure 6.7
Distortion of the fate map in bicoid *mutant embryos as revealed by fushi tarazu protein expression. The pattern of* fushi tarazu *stripes is shifted anteriorly, and stripes 2 and 3 are fused. After Mlodzik et al., The influence on the blastoderm fate map of maternal-effect genes that affect the antero-posterior pattern in* Drosophila, Genes and Development *1 (1987): 603–14.*

phone system. If you want to wire the city of Paris, you only need ten digits (0 through 9) and you can compose ten million phone numbers consisting of seven digits each. In analogy you would need only a limited number of transcription factors (encoded by regulatory genes) to interconnect a large number of target genes. Each target gene has a combination of binding sites like a phone number. Of course, the analogy is not perfect because, for example, the sequence of the binding sites (digits) may be changed, but it reflects the general idea of combinatorial interactions rather well. If each gene were regulated independently, there would be no coordination and development would be chaotic. Many genes therefore are highly pleiotropic, meaning that they exert a multitude of effects by interacting with many other genes or their gene products.

The downstream target genes of *fushi tarazu* can be identified by analogous experiments in which the expression of the downstream gene is analyzed in a normal (*fushi tarazu*$^+$) and mutant (*fushi tarazu*$^-$) background. For example, *engrailed,* which basically is expressed in fourteen stripes (one per segment), expresses seven stripes only in *fushi tarazu*$^-$ embryos (see fig. 6.6). Because the seven *engrailed* stripes are missing in these regions where *fushi tarazu* is normally expressed, we can conclude that *fushi tarazu* directly or indirectly activates *engrailed.* In summary, *fushi tarazu* is regulated by genes upstream in the regulatory cascade, it autoregulates itself, and it in turn regulates downstream target genes.

With respect to the regulatory cascade, the classification of segmentation genes into gap, pair-rule, and segment polarity genes has greatly clarified the picture. The body plan of the embryo is gradually established by subdividing the embryo on the basis of the spatial coordinates provided by the morphogenetic gradients first into large domains comprising several segments and then progressively into smaller stripes until finally each cell is specified (pl. 3).

The pioneering work on the gap genes was carried out by Herbert Jäckle and his collaborators. Some scientists have green fingers, but Jäckle is different—he has zinc fingers: every gene that he first touches seems to have zinc finger motifs, and this is true for most of the gap genes. Zinc fingers are DNA-binding domains of gene-regulatory proteins in which zinc is bound by invariant pairs of cysteines and histidines embedded in a fingerlike structure. Zinc finger proteins can serve both as DNA and RNA binding proteins. The expression pattern of gap genes, as for example *Krüppel* (*Kr*), are

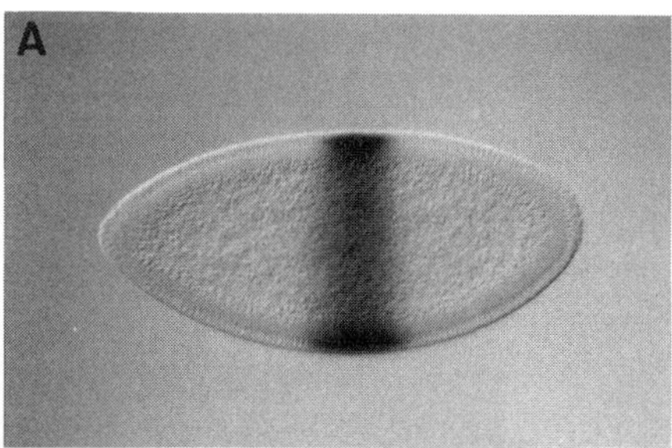

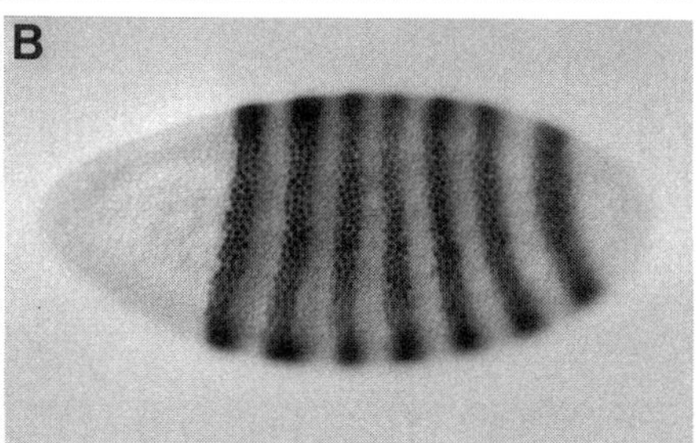

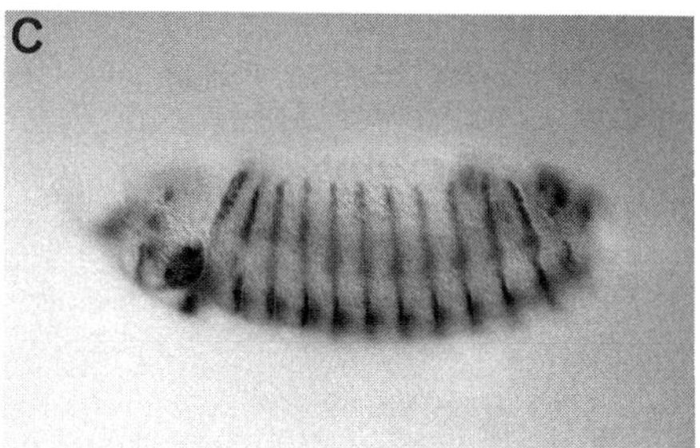

Figure 6.8
Hierarchy of the segmentation genes. Progressive subdivision of the embryo into broad domains (A), (para-)segments (B), and compartments (C): (A) expression of gap gene Krüppel *at the preblastoderm stage; (B) expression of the pair-rule gene* fushi tarazu *at the preblastoderm stage; (C) expression of the segment-polarity gene* engrailed *at a later stage of segmentation. Courtesy of S. Flister.*

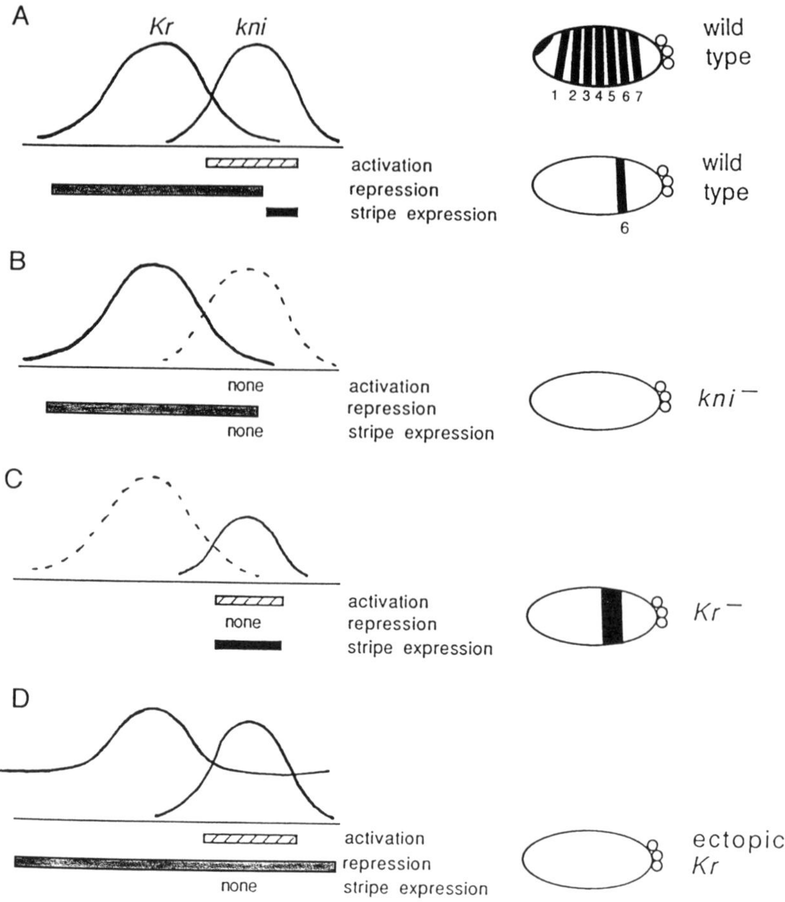

Figure 6.9
Formation of individual stripes: hairy *stripe number 6. From M. J. Pankratz and H. Jäckle, Making stripes in the* Drosophila *embryo,* Trends in Genetics *6 (1990): 287–92.*

affected in maternal-coordinate mutants like *bicoid,* and also in other gap mutants like *hunchback* and *knirps* (*kni*), but their expression is not changed in pair-rule and segment polarity mutants. Because pair-rule gene expression is altered in maternal-coordinate, gap, and other pair-rule mutants but unaffected by segment polarity mutants, the developmental hierarchy goes from maternal-coordinate to gap to pair-rule to segment-polarity genes (fig. 6.8), which is also consistent with the time course of their expression. Furthermore, there are cross-regulatory interactions among genes at the same level of the hierarchy as described above for *fushi tarazu* and *hairy.*

The next question we can now address is how to form a stripe. If an engineer were asked to convert a graded pattern into a periodic pattern of stripes he or she would simply convert the gradient into a wave function that would simultaneously generate all stripes. But this is not how nature operates: each stripe in the *Drosophila* embryo is regulated differently, by a different combination of genes. This is nicely illustrated with stripe number six of *hairy* expression (fig. 6.9). In the normal (wild-type) embryo (fig. 6.9A) the two gap genes *Krüppel* and *knirps* are expressed in two broad, bell-shaped, partially overlapping domains. *knirps* is an activator of *hairy, Krüppel* a repressor. Therefore, the sixth *hairy* stripe is formed only in the area where the two gene expression domains do not overlap. If one hand, *knirps* is removed by mutation, the activator is missing and no sixth stripe is formed (fig. 6.9B). If, however, *Krüppel,* the repressor, is removed, the sixth stripe expands in the anterior direction (fig. 6.9C). Finally, when the *Krüppel* repressor is artificially expressed all over the embryo, no sixth *hairy* stripe is formed (fig. 6.9D). The other *hairy* stripes are controlled by different combinations of genes. The formation of the stripes is the product of long evolutionary processes that resemble tinkering much more than straightforward design. The control regions of pair-rule genes like *hairy* and *even-skipped* (*eve*) include a modular array of "stripe elements." As most clearly shown by Michael Levine and his collaborators, each stripe element contains a specific set of activator and repressor binding sites. The best-studied example is the second stripe of *even-skipped* expression: The stripe two element contains binding sites for the activator proteins bicoid and hunchback, as well as for the repressors Giant and Krüppel, whereas the other stripes are generated by different combinations of regulatory proteins.

The pair-rule stripes that are one parasegment wide are further subdivided by the segment-polarity genes. At late blastoderm and early gastrula stages (later than *fushi tarazu*), another homeobox gene, *engrailed,* becomes

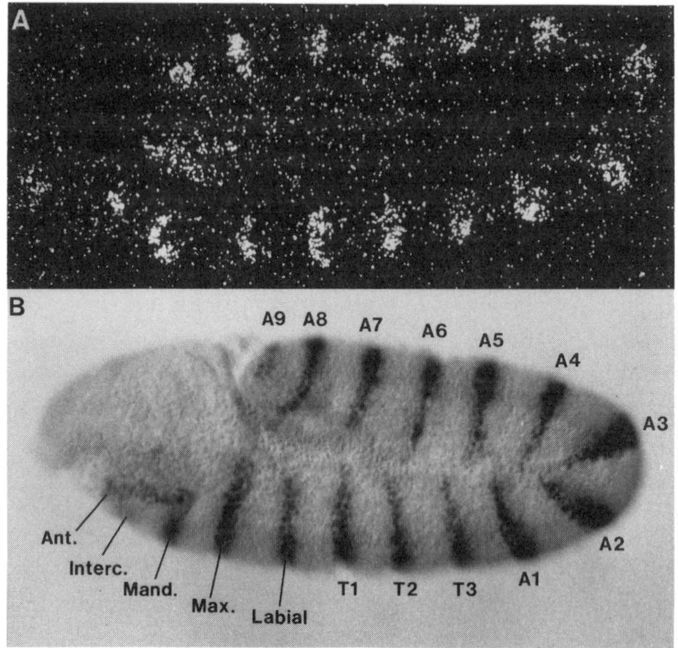

Figure 6.10
Expression of pattern of engrailed *RNA and protein: (A) detection of* engrailed *messenger RNA by in situ hybridization; (B) detection of engrailed protein by antibody staining. Courtesy of A. Fjose and S. Flister.*

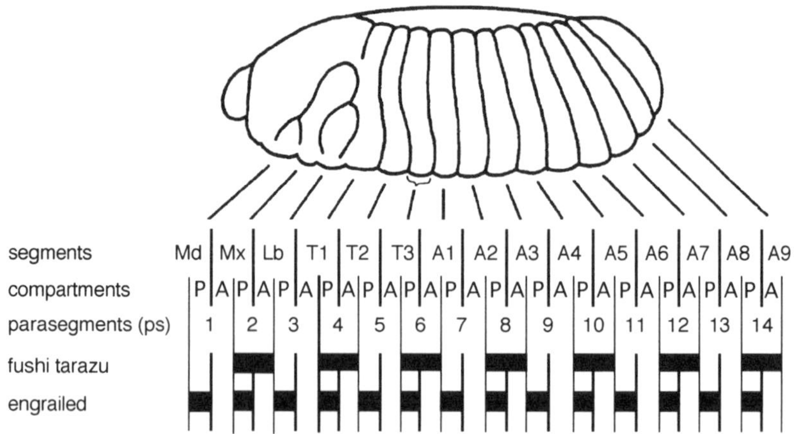

Figure 6.11
Expression of fushi tarazu *and* engrailed *in segments and parasegments*

expressed. It was cloned by Anders Fjose in my laboratory on the basis of homeobox homology and in Tom Kornberg's group by chromosome walking. Earlier genetic experiments by Gines Morata, Peter Lawrence, and Antonio Garcia-Bellido had indicated that each segment can be subdivided into an anterior and a posterior compartment and that *engrailed* is required in the posterior compartment. Our in situ hybridization experiments supported this conclusion strongly. The first picture Fjose obtained is depicted in figure 6.10. When the pattern of expression of *fushi tarazu* and *engrailed* are compared at a stage when the segmentation pattern of the embryo is clearly visible, it becomes apparent that *fushi tarazu* is expressed across the segmental boundaries in stripes that correspond to the even-numbered parasegments from parasegments two to fourteen, whereas *engrailed* expression is confined to the posterior compartment of each body segment (fig. 6.11). Thus, starting from morphogenetic gradients the body plan is subdivided first by the gap genes into broad domains, then by the pair-rule genes into segmentally repeated units, and last by segment-polarity genes into compartments (half-segments) in a stepwise manner.

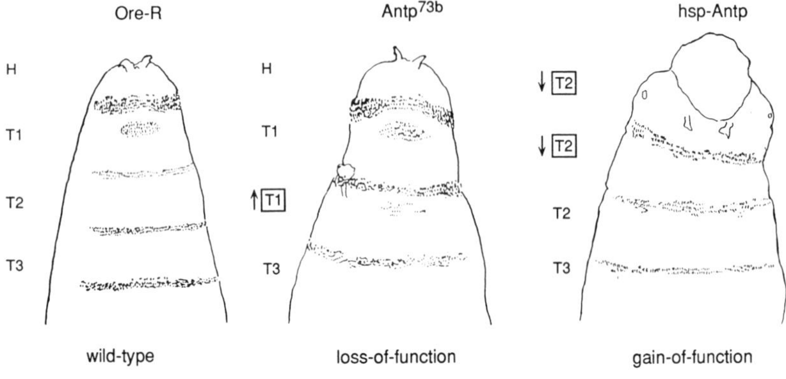

Figure 7.1
Homeotic transformations in Antennapedia *mutants. The wild-type larva shows the normal pattern of segmentation into head (H) and thoracic segments (T1, T2, and T3). In loss-of-function mutants (*$\text{Antp}^{73b}/\text{Antp}^{73b}$*), T2 is transformed into T1 (anterior transformation). In gain-of-function mutants (hsp-*Antp*), H and T1 are transformed in T2 segments (posterior transformation). Courtesy of T. Kaufman.*

~

7
The Master Control Genes

In January 1986, at *Nature*'s first conference in Japan, prominent scientists from throughout the world were invited to report on the latest progress in molecular biology. In the program was this statement: "The conference will be honored by the presence of His Imperial Highness Prince Hitachi, who will give an opening address." In a separate letter, the speakers were informed that His Highness would give a reception before the opening ceremony and would meet the speakers at the reception hall of the large hotel in which the conference was being held. Accordingly, we assembled in the hall. In due time the prince appeared with his staff, and we were asked to follow him into a room where tea would be served. He took a seat on a large sofa with a couple of cushions and his secretary informed us that he wished to have a brief private conversation with each of us in turn. Now, when I have to present a lecture in front of a large audience I no longer get nervous, but on this occasion I became very anxious because I had no idea of what I should discuss with the prince. I knew too little about Japanese politics and cultural life to hold an interesting conversation with the prince, and after all you could not talk about the weather. And so I grew increasingly nervous. Sidney Brenner was introduced to the prince just before me, and soon the two were laughing as if they were old pals. But this did not relieve my anxiety, because Sidney is always witty and often in the mood for joking. Then it was my turn. The secretary guided me over to meet the prince and introduced me by my name. I bowed as well as I could, but it must have looked pretty awkward. (The British are much better at that because they still have a queen, but the Swiss fought the Habsburgs back in the thirteenth century and have not had a king or an emperor since.) In any case, Prince Hitachi asked me to sit down next to him, and before I could think of anything to say, he asked, "What's new about the homeobox?" I must have made the most astonished face in my entire life, but his question certainly broke the ice. I had been told before that the prince trained as a scientist, but I certainly did not expect him to know about my work. I gave him a rundown of our latest results, which he followed with great interest. He spoke perfect English, and when I complimented him on his mastery of the language, he dismissed my remark by saying that he had been educated at Cambridge. After the reception, he proceeded into the lecture hall, followed by his staff and the speakers. To my surprise, the audience stood and waited respectfully until first Prince Hitachi and then we speakers had taken our seats. I don't remember too much about the opening ceremony, but I do remember that the Japanese audience was most impressed by my discovery of the eight-legged butterflies in front of the Buddha in Nara, as I mentioned in Chapter 2 (see fig. 2.1).

This brings us back to the homeotic genes and their function in development. In *Drosophila,* the establishment of the body plan occurs in four conceptual steps (pl. 3). The first step involves the localization of a cytoplasmic determinant, in this case the messenger RNA of the maternal coordinate gene *bicoid* in the unfertilized egg. On fertilization the localized RNA becomes translated into bicoid protein, which spreads by diffusion and forms a morphogenetic gradient. The bicoid protein concentration provides positional information to the nuclei in which it is taken up. The third step involves the conversion of the graded pattern into a repetitive pattern of stripes under the control of the segmentation genes, as discussed in Chapter 6. Finally, the repetitive pattern is converted into a sequential pattern; the segments differentiate and each acquires its own identity (positional information along the antero-posterior axis). Segmental identity and the proper sequence of the segments are specified by the homeotic genes. Homeobox genes are involved at each of these conceptual steps, but the homeotic genes in the strict sense of the word are mainly involved in this fourth step, in specifying segmental identity. Furthermore, the order of the segments is determined by mutual interactions between homeotic genes. In addition to this region-specific action, homeotic genes that determine organ-specificity and tissue- or cell-type specificity have been isolated. In general, these genes determine cell fate in a particular region of the body, within a body segment, in an organ, or in individual cells.

As a paradigm let us consider *Antennapedia,* the gene I have studied most intensively. One reason I chose *Antennapedia* is that it is a switch gene or, as Antonio Garcia-Bellido later called it, a selector gene. It causes a switch, or selects, between two alternative developmental pathways. Such genes are much more likely to have a regulatory function than genes that block a developmental pathway when they are mutated. Yet this argument is only partially correct, as we shall see. The developmental switch in *Antennapedia* can occur in both directions, depending on the nature of the mutation: gain-of-function mutants like *Nasobemia* are dominant and transform the antennae into middle legs, whereas loss-of-function mutants are recessive and lead to a transformation of middle legs toward antennae. Because the recessive loss-of-function mutants are lethal at the embryonic or larval stages, this experiment had to be carried out in genetic mosaic flies. As first shown by Gary Struhl, clones of homozygous *Antennapedia*$^-$ mutant cells, when located on the middle legs, form antennal rather than leg structures, whereas clones arising in other parts of the fly differentiate normally. The opposite direction of the transformations caused by gain- and loss-of-function mutants can also be detected in the embryos (fig. 7.1). The late embryonic segmentation pattern is characterized by specific denticle belts that allow the assignment of the thoracic and abdominal segments. In the wild-type embryo the first thoracic segment, T1, is characterized by a special beardlike structure that is absent from T2 to T3. In loss-of-function mutant embryos T2 is partially transformed into T1 (T2→T1) with an additional beard, so that it lacks a normal T2 segment. In strong gain-of-function mutants the opposite transformation T1→T2 (absence of the beard) is observed, and even the more anterior head segments are converted into T2. The absence of a T2 segment in loss-of-function mutants and the formation of additional T2 segments in gain-of-function mutants suggested that *Antennapedia* specifies T2—that is, the middle leg and not the antenna.

This hypothesis is consistent with the early pattern of *Antennapedia* expression in the embryo. *Antennapedia* is transcribed at the blastoderm or early gastrula stage as a single stripe corresponding to the second thoracic segment, and no expression can be detected in the antenna region of the head (pl. 2). At later stages, however, expression can also be detected more posteriorly, even though the Antennapedia protein primarily accumulates in the three thoracic segments, in both the epidermis and the ventral nervous system. In the ventral nerve cord it accumulates in the posterior compartment of T1 and the anterior compartment of T2 (parasegment 4), leaving a gap in posterior T2, and being expressed again in the anterior compartment of T3. In larvae it is expressed similarly in all three thoracic ganglia of the ventral nervous system and in all thoracic imaginal discs, but not in the eye-antennal disc. However, the expression pattern is less clear-cut than one might expect: only the dorsal prothoracic disc (T1) expresses *Antennapedia* uniformly in all its cells, whereas the Antennapedia protein is expressed mainly in the posterior compartments of the foreleg disc and the anterior compartment of the middle and hindleg discs. Furthermore, expression in the wing and haltere discs is confined to those parts that give rise to the notum (the dorsal thorax), and little expression is detected in the future wing and haltere cells. If *Antennapedia* were involved only in specifying T2, one might have expected its expression to be confined to the middle leg and the wing discs only. Further, it should be expressed in all cells of these discs, but neither of these expectations were fulfilled. This is an important point that we shall consider again below.

The key to understanding how *Antennapedia* functions was provided by the dominant gain-of-function mutants that induce antenna-to-leg transformations. But what was their molecular basis? How could a chromosomal inversion, with one breakpoint inside the gene and one outside, give rise to a dominant gain-of-function phenotype? This problem gave me numerous sleepless nights. The term *gain-of-function mutant* was coined by Ed Lewis, who had discovered a dominant mutant counterpart to the recessive *bithorax* mutations. He named it *Contrabithorax*, since it gave the opposite phenotype—namely, four halteres instead of four wings as produced by *bithorax*, whereas the normal wild type fly has one pair of wings and one pair of halteres. The *bithorax* mutations were certainly caused by a loss-of-function, because deletions of the gene have the same phenotype as the original *bithorax* mutation. Not only does *Contrabithorax* have the opposite phenotype, but flies carrying one *Contrabithorax* mutation and two wild-type genes also show that mutant phenotype. Lewis therefore considered *Contrabithorax* to be a gain-of-function mutation, but the molecular basis of the mutation remained enigmatic. When David Hogness's group had cloned *bithorax*, they also looked at *Contrabithorax* and found that it was associated with a complicated chromosome rearrangement (a transposition plus an inversion of a DNA segment) that was impossible to interpret with certainty. This was not much help, and so I asked Stephan Schneuwly, one of my very able graduate students, to examine *Antennapedia*^{73b}, an inversion mutant with a dominant gain-of-function phenotype, in order to find its molecular lesion. Structural analysis of the *Antennapedia* gene had shown that in all of the known dominant gain-of-function mutants the protein-coding region is not altered but instead separated from the promoter by the inversion breakpoint. The dominant mutant phenotype therefore is presumably due not to an altered gene product but rather to the altered expression of the normal Antennapedia protein. At least for *Antennapedia*^{73b}, Schneuwly was able to show that the chromosomal inversion results in the fusion of an upstream control region and RNA coding sequences of a foreign gene to the Antennapedia protein-coding region. This rearranged gene gives rise to a fusion transcript and is controlled by the foreign promoter. Presumably for this reason it was expressed ectopically—that is, in the antennal disc rather than the leg disc. Because at least eight independent inversion mutants were known to produce antenna-to-leg transformations I thought unlikely that all these inversions were due to fusions of the Antennapedia protein-coding region to antenna-specific promoters. It seemed that any promoter that caused ectopic *Antennapedia* expression in the antennal disc at the appropriate time in development could induce antenna-to-leg transformations. Such a promoter might also be active in other tissues, but the effect would be limited to the antennae if the protein were expressed at high enough levels at the critical time when the antennae become determined.

This led me to the idea of constructing an artificial *Antennapedia* fusion gene to test this hypothesis. I called a group meeting and presented my plan of fusing a heat-shock promoter to the protein coding of *Antennapedia.* The heat-shock promoter can be activated at will by exposing the fly briefly to higher temperature so that the synthesis of Antennapedia protein can be induced at any stage of development in all the cells of the fly. Because the *Antennapedia* gene was much too large to handle, I proposed to use the cDNA only (lacking the introns) even though this had not been tried before. The prediction was that a heat shock during the critical time would induce antenna-to-leg transformations much as in natural dominant gain-of-function mutants. The group's response ranged from enthusiasm to skepticism. Stephan Schneuwly was quite excited about the idea and willing to try the experiment, whereas Roman Klemenz was more skeptical, though willing to construct the heat-shock vector because he was working on the regulation of heat-shock genes anyway. Klemenz designed a heat-shock vector that with some minor improvements has been extremely useful for many years. Schneuwly inserted an *Antennapedia* cDNA with a complete protein coding region into this heat-shock vector and obtained transgenic flies carrying this fusion construct. In these transgenic flies, the synthesis of Antennapedia protein can be induced in all cells by a heat shock—a raise in temperature from 25°C to 37°C for two hours or more. By antibody staining the Antennapedia protein could be detected in the nuclei of all the cells of heat-shocked embryos, and particularly in the antennal discs of heat-shocked larvae. However, in the first experiments the heat-shocked flies all had normal antennae. It was difficult to determine the critical time period for heat-shock induction. However, a week later Stephan observed three bristles on the antenna of a heat-shocked fly, which I identified as leg bristles on the basis of the bracts at their base, and I became convinced that the experiment would succeed. After another few weeks of hard work, Stephan was able to induce large antennal legs with claws and all the characteristics of middle legs (pl. 4). This was the first time that the body plan of an animal had been experimentally altered in a predicted way, and we proudly called it redesigning the body plan of the fruit fly. These experiments demonstrated that the assignment of this DNA segment to the *Antennapedia* gene was correct and showed conclusively that *Antennapedia* is a master control gene for leg morphogenesis. Turning on a single master control switch activates the leg developmental pathway, which involves several hundred or thousand genes.

What about the dorsal part of the second thoracic segment? Besides the antenna-to-leg transformations, the heat-shocked transgenic flies also showed head-to-thorax transformations (pl. 4). These correspond to another type of dominant gain-of-function mutations, called *Cephalothorax* (that is associated with a complicated chromosomal rearrangement, with one breakpoint at the *Antennapedia* locus. The thoracic structures are derived from those areas of the wing disc that in the wild-type wing disc express *Antennapedia.* Loss-of-function mutations ($Antp^{PW}/Antp^{PW}$) show the opposite transformation of this area from thorax to head. However, no head-to-wing transformations were observed in the heat-shocked transgenic flies, and there are also no gain-of-function mutants known at the *Antennapedia* locus that cause such transformations. Another gene, however, called *ophthalmoptera* (*opht*), causes eye-to-wing transformations. It is located on a different chromosome and remains to be cloned and analyzed.

When *Antennapedia* is expressed in the eye imaginal disc, it induces cell death (apoptosis), leading to a reduction in the number of facets in the eye, rather than eye-to-wing transformations. If, however, we prevent cell death in the eye imaginal disc by a genetic trick, the *Antennapedia*-expressing cells form little wings in the eye region. This indicates that *Antennapedia* specifies not only the leg but the entire second thoracic segment, including the dorsal thorax and the wing.

Antenna-to-leg transformations have since been found by Richard Mann also to occur in heat shock–*Ultrabithorax* fusion constructs, but they appear to be hind legs rather than middle legs. Heat shock–*Sex combs reduced* fusions made by Peter LeMotte in my laboratory led to partial transformations of the aristae at the tip of the antennae into tarsal structures, and we do not know whether they correspond to first legs. Because more than one homeotic gene is expressed per thoracic segment, it seems that the combinatorial action of several genes is required to specify an entire segment.

These experiments involve another complication: the ectopically expressed homeotic gene has to compete with the "resident" homeotic gene activity. The antenna disc seems to be a relatively weak spot in the genetic circuitry that can be completed successfully by *Antennapedia*, whereas the more posterior segments are harder to transform. Both competitive interactions for the target genes and mutual interactions among the homeotic genes are important, but we must decipher more of the genetic circuits controlling development before we can fully understand the principles underlying the specification of the body plan in both molecular and evolutionary terms.

An important part in this undertaking has been played by the genes of the Bithorax complex. Ed Lewis's model originally assumed that one homeotic gene specified each segment and the sophisticated genetic mapping data indicated that they were arranged in the same sequence along the chromosome as they were expressed along the antero-posterior body axis. However, when Gines Morata and his colleagues saturated the Bithorax complex with lethal mutations, they found only three genes: *Ultrabithorax*, *abdominal-A*, and *Abdominal-B*. This contradiction was resolved by the cloning of the Bithorax complex. In David Hogness's laboratory Welcome Bender, Pierre Spierer, and others performed an extensive chromosomal walk in which they first cloned the *Ultrabithorax* gene, which was shown by both my group and Matthew Scott's group to contain a homeobox. Using the homeobox as a probe we isolated *abdominal-A* and later found a third homeobox corresponding to *Abdominal-B*. François Karch and Welcome Bender's completion of the chromosome walk showed that there were indeed only three

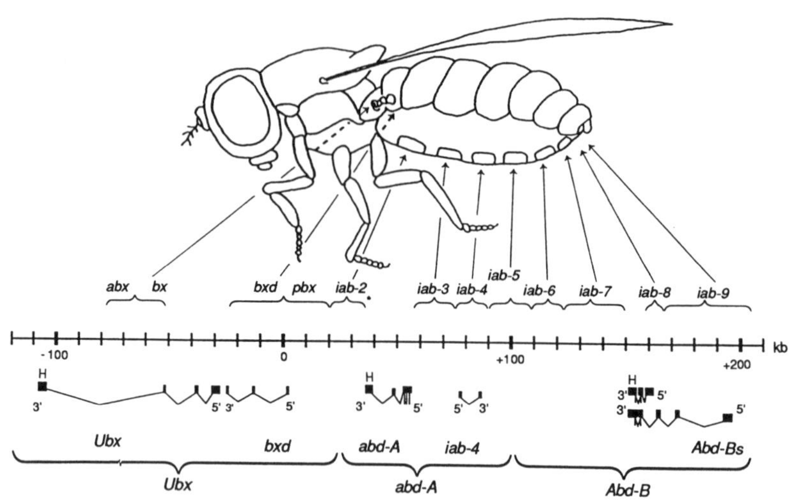

Figure 7.2
Genomic organization of the Bithorax complex. The three genes Ultrabithorax *(*Ubx*),* abdominal-A *(*abd-A*), and* Abdominal-B *(*Abd-B*) and their cis-regulatory regions* abx *to* iab-9 *are arranged in the same order along the chromosome, represented as 330 kilo base pairs (kb) of DNA, as they are expressed along the antero-posterior body axis in the fly (arrows). After E. B. Lewis, Clusters of master control genes regulate the development of higher organisms,* Journal of the American Medical Association *267 (1992): 1524–31.*

protein-coding genes, *Ultrabithorax, abdominal-A,* and *Abdominal-B,* as proposed by Gines Morata, and that each had a homeobox. The other genes identified by Ed Lewis were shown to be regulatory regions that seemed to function as segment-specific enhancers. Several of these regulatory regions were shown to be transcribed into RNA, however, and there is some lingering doubt that they might encode RNAs with a regulatory function. However, the colinearity rule applies even in this modified version of the Lewis model (fig. 7.2).

On the basis of his genetic analysis, Thomas Kaufman proposed that the Antennapedia complex represents a gene complex specifying the anterior thoracic and the head segments in the same way that the Bithorax complex determines the posterior thoracic and the abdominal segments. The picture was confused at first by the presence of nonhomeotic genes in the Antennapedia complex, but gene cloning eventually proved Kaufman to be right. The anterior-most expressed *labial* (*lab*) gene is followed by *proboscipedia* (*pb*), *Deformed, Sex combs reduced,* and *Antennapedia,* which are expressed in progressively more posterior segments (fig. 7.3). Their expression domains are roughly indicated in figure 7.4. They generally comprise more than one segment and overlap even though two genes may not be expressed in the same cells. This is consistent with a combinatorial model in which more than one gene contributes to the specification of one segment.

In the beetle *Tribolium castaneum,* the homeotic genes of the Antennapedia and Bithorax complexes form a single cluster. Because a homologous cluster is found even in mammals, as discussed later, the cluster must have arisen early in evolution, before vertebrates and invertebrates separated, and must have formed the basis of the metazoan body plan.

The mutual interactions between homeotic genes were again first demonstrated with *Antennapedia* and the genes of the Bithorax complex. Ernst Hafen and Michael Levine used the method of in situ hybridization that they had developed to analyze the expression of *Antennapedia* in wild-type and *bithorax* mutant embryos. Mutant embryos lacking the *Ultrabithorax* gene show a transformation of the third thoracic (T3) and first abdominal segment (A1) into second thoracic segments (T2) and therefore have three consecutive T2 segments rather than one as in the wild type. Deletion of all three genes of the Bithorax complex (*Ultrabithorax, abdominal-A,* and *Abdominal-B*) leads to the transformation of all posterior seg-

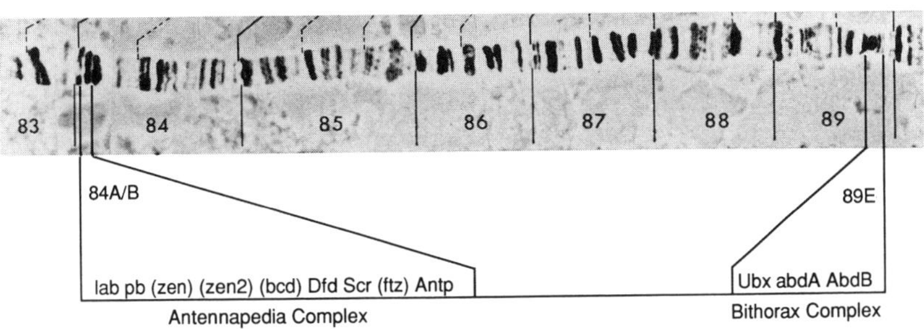

Figure 7.3
Giant polytene chromosome 3 (right arm), indicating the localization of the Antennapedia and Bithorax complexes in sections 84A/B and 89E, respectively. Homeobox genes not involved in antero-posterior specification of the body plan are shown in parentheses.
Abbreviations: lab labial, pb proboscipedia, zen zerknüllt, bcd bicoid, Dfd Deformed, Scr Sex combs reduced, ftz fushi tarazu, Antp Antennapedia, Ubx Ultrabithorax, abdA abdominal-A, AbdB Abdominal-B.

ments into T2 (see Deficiency P9 in fig. 2.5). In situ hybridization with an *Antennapedia* probe to wild-type (fig. 7.5), *Ultrabithorax*$^-$, and P9 embryos reveal that *Antennapedia* expression is strictly associated with the formation of the second thoracic segment (fig. 7.6). This implies that *Antennapedia* is repressed in the posterior body segments of the normal embryo by the genes of the Bithorax complex. Removal of these genes in mutant embryos derepresses *Antennapedia* in the posterior segments and thereby causes their transformation into second thoracic segments. Because the Ultrabithorax protein binds in vitro to the regulatory region of *Antennapedia*, it has been proposed that the Ultrabithorax protein directly represses *Antennapedia* transcription in the normal embryo. But there is also evidence for a competition between *Antennapedia* and *Ultrabithorax* for their target genes, which in the posterior segments is decided in favor of *Ultrabithorax* and the other posteriorly expressed genes (*abdominal-A* and *Abdominal-B*). Nevertheless, in normal embryos *Antennapedia* is also expressed in some cells of the segments posterior to T2, and it apparently also contributes to the differentia-

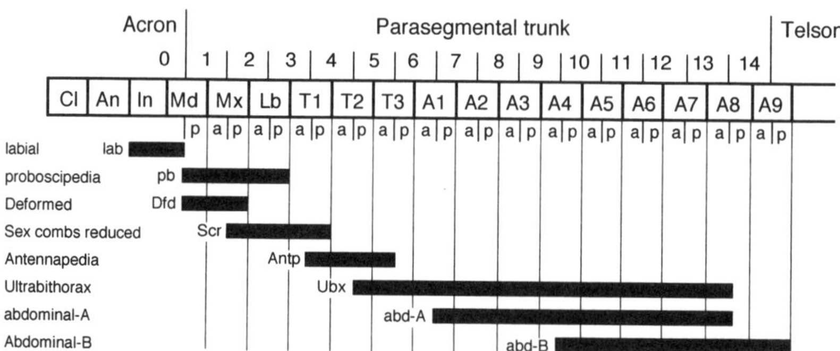

Figure 7.4
Expression pattern of homeotic genes along the antero-posterior body axis of Drosophila. *The anterior is to the left. The acron (anterior head region) consists of three fused head segments, clypeolabrum (Cl), antenna (An), and intercalary (In), and the parasegmental trunk comprises the 14 parasegments that form the mandibula (Md), maxilla (Mx), labium (Lb), the thoracic segments (T1–3), and the abdominal segments (A1–9), terminating with the telson. The expression patterns are simplified, especially with respect to the posterior expression boundaries. From D. Bachiller, A. Macias, D. Duboule, and G. Morata, Conservation of a functional hierarchy between mammalian and insect* Hox/HOM *genes,* EMBO Journal *(1994): 1930–41.*

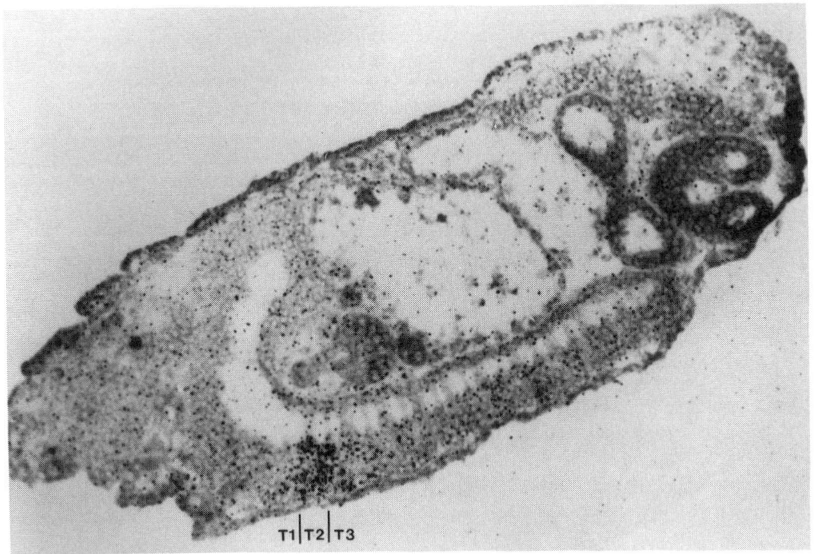

Figure 7.5
Detection of Antennapedia *transcripts in the three thoracic segments. After E. Hafen, M. Levine, R. L. Garber, and W. J. Gehring, An improved* in situ *hybridization method for the detection of cellular RNAs in* Drosophila *tissue sections and its application for localizing transcripts of the homeotic* Antennapedia *gene complex,* EMBO Journal 2 *(1983): 617–23.*

tion of the posterior segments, though to a lesser extent. A given body segment therefore appears to require a certain combination of homeotic genes for proper differentiation, as postulated by the Lewis model.

A number of homeobox genes have been found outside the clusters, however. They either originated in the cluster and were subsequently translocated or never formed part of the cluster. One of these genes, *eyeless* (*ey*), has become the paradigm for a master control gene. In addition to a homeobox, *eyeless* also contains a paired box encoding a second type of DNA binding domain. *eyeless* corresponds to the mutation *Small eye* in mice and to *Aniridia* in humans. At first glance, this gene is not necessarily a switch gene, because its recessive loss-of-function phenotype in addition to other defects prevents eye formation and leads to a reduction or absence of the eyes. As shown in nematodes, however, cell death can be a programmed cell fate. In experiments similar to those described above for *Antennapedia,* the *eyeless* gene can be ectopically expressed in the wing, leg, or antenna discs and

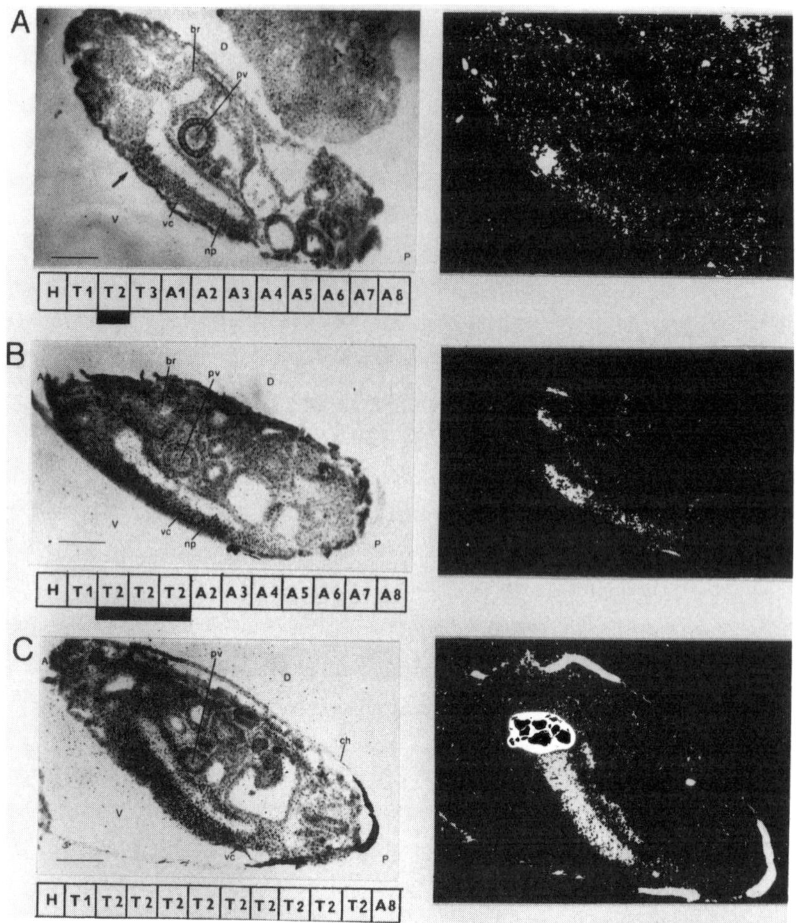

Figure 7.6
Expression of Antennapedia *transcripts in deletions of the Bithorax complex: (A) in the wild-type embryo the expression is strongest in the mesothorax (T2); (B) in a deletion of* Ultrabithorax, *T3 and A1 are transformed to T2 and the expression of* Antennapedia *is broadened to all three T2 segments; (C) in a deletion of the entire Bithorax complex, T3–A7 are transformed to T2 segments and the expression domain of* Antennapedia *expands correspondingly.* Abbreviations: A *anterior,* br *brain,* D *dorsal,* np *neuropile (nerve fibers),* P *posterior,* pv *proventriculus,* V *ventral,* vc *ventral chord. From E. Hafen, M. Levine, and W. J. Gehring, Regulation of* Antennapedia *transcript distribution by the Bithorax complex in* Drosophila, Nature *307 (1984): 287–89.*

induce the formation of ectopic eyes on the wings, legs, and antennae of the transgenic fly. These ectopic eyes contain functional photoreceptors and obviously represent real eyes, not just eyelike structures. Apparently, then, a single master control gene can switch on the eye developmental program, which involves more than 2,500 other genes needed for eye morphogenesis. These findings open a new chapter of developmental biology, and their evolutionary implications are discussed in the final chapter of this book.

Figure 8.1
Egg-eating snake. Courtesy of K. Wüthrich.

~

8
The Molecular Basis

At one of the annual meetings of the Swiss Societies for Experimental Biology I heard a plenary lecture by Kurt Wüthrich that was to have a considerable impact on my future research. Wüthrich began by showing some spectacular slides of egg-eating snakes, which crack the problem of the secret contents of the egg in their own fashion (fig. 8.1). This immediately showed me that he had not only a good sense of humor but also broad interests ranging from physics and chemistry to biology. In the course of his talk he demonstrated his sophisticated methods for deriving the three-dimensional structure of small protein molecules from nuclear magnetic resonance (NMR) spectra (fig. 8.2). It seemed a miracle that he could make sense of this maze of thousands of peaks, but a comparison with the data obtained by Robert Huber using X-ray crystallography showed that the results were essentially the same.

As a biologist, I was particularly interested in Kurt Wüthrich's NMR data, because I intuitively felt that a structure obtained in solution is closer to life than one obtained in crystal form. I had debated this point several times with my friends from biophysical chemistry and structural biology at

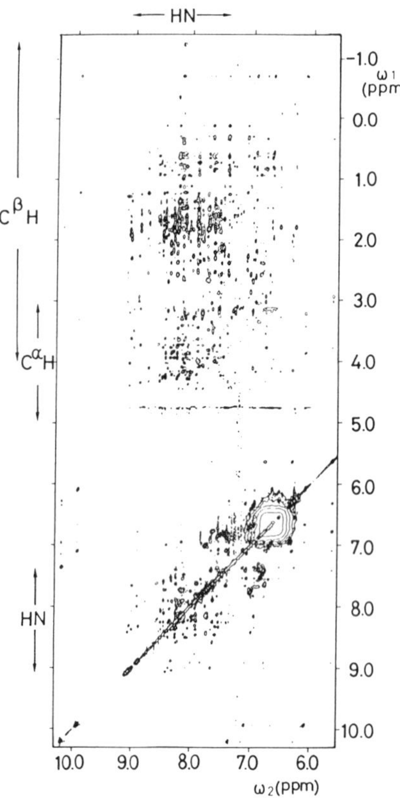

Figure 8.2
Nuclear Overhauser enhancement spectrum of the Antennapedia *homeodomain polypeptide recorded at 600 megahertz. Spectral region with resonances from all amide protons (NH) along the* ω_1 *and the complete spectrum for the amide protons,* α *and* β *carbons (*$C^{\alpha}H$ *and* $C^{\beta}H$*), along the* ω_2 *axis. From G. Otting, Y.-Q. Qian, M. Müller, M. Affolter, W. J. Gehring, and K. Wüthrich, Secondary structure determination for the* Antennapedia *homeodomain by nuclear magnetic resonance and evidence for a helix-turn-helix motif,* EMBO Journal 7 *(1988): 4305–9.*

the Biozentrum, who dismissed my criticism, but I was not entirely convinced by their arguments. To me the crystal structures were too static and did not reveal the more dynamic aspects of the molecule. Because I realized that we could not understand the function of the homeodomain without knowing its structure, I called Kurt Wüthrich and asked him over the phone whether he would be interested in a collaboration on the homeodomain and how much protein he would need to determine its structure. After roughly calculating the molecular weight of the isolated homeodomain, he answered in his slow Bernese accent that some fifty milligrams might do, but what was most important was that the protein had to be very pure. I almost had a heart attack, because I was used to thinking in terms of micrograms rather than milligrams, and while I was trying to recover from this shock, he asked me whether I still was on the phone. Having recovered my courage, I said that we would try to isolate this large amount and that I would call him back.

After this conversation, I talked to my bravest graduate student, Martin Müller, telling him that this was an excellent project, and he accepted the challenge even though I told him that milligram quantities were required. At that time, 1987, it was difficult to synthesize large amounts of homeodomain proteins in bacteria because these DNA-binding molecules are highly toxic for the bacterial cells. But William Studier had developed a set of vectors based on bacteriophage T7, which allowed us to overcome this problem. Müller inserted the *Antennapedia* homeobox into this vector, and *E. coli* collaborated very nicely and synthesized large amounts of homeodomain protein for us. He worked out a simple purification procedure involving four basic steps, and succeeded in purifying milligram quantities fairly easily. His first preparation was eleven milligrams and we carried it to Zurich—it was so precious that I didn't want to risk its getting lost in the mail. Kurt Wüthrich and his collaborators, who can detect small traces of contaminating molecules in their spectra, judged the sample as highly pure and found that the amount was enough for a complete structure determination.

With substantial amounts of homeodomain protein at hand and the possibility of synthesizing short oligonucleotides corresponding to its DNA binding site, we could now study the DNA-protein interactions biochemically, which brought me back to DNA binding studies, as I had carried them out as a postdoctoral fellow at Yale. Fortunately, methods had improved considerably since then, and it had become possible to study protein-DNA interactions in much greater detail. In particular methods for "footprinting," methylation and ethylation interference, and gel mobility shift assays had been added to the repertoire. These experiments were carried out mainly by Markus Affolter and Tony Percival-Smith in the laboratory and provided important complementary information to the structural studies carried out in Zurich. Footprinting basically involves the digestion of a DNA-protein complex with DNAse; the free DNA is readily digested, whereas the DNA segment that is bound to the protein is protected from degradation, and so the protein leaves a "footprint" on the DNA. Similarly, the methylation or ethylation of certain bases may interfere with the binding of the protein and alter the pattern of accessibility for the cleavage by DNAse. Finally, the binding of a protein molecule to an oligonucleotide alters the molecular weight of the oligonucleotide and its migration in an electrical field during gel electrophoresis. Using these techniques, Affolter and Percival-Smith could identify DNA binding sites in the regulatory region of the *engrailed* gene, to which an intact fushi tarazu protein, a truncated *Antennapedia* protein, and the isolated *Antennapedia* homeodomain bound with high affinity. Later we could show that these binding sites of fourteen base pairs minimally are functional in vivo. Contrary to our expectations based on bacterial repressors, which bind as dimers or tetramers, the homeodomain binds, at least in vitro, as a monomer with high affinity to its binding site. The equilibrium dissociation constant is 1.6×10^{-9} M (Molar) and the homeodomain-DNA complex is very stable, with a half-life of approximately ninety minutes. These values are important because they enabled us to study the homeodomain-DNA complex by NMR spectroscopy as well.

In a short time Kurt Wüthrich and his collaborators Yan-qiu Qian, Gottfried Otting, and Martin Billeter solved the three-dimensional structure of the *Antennapedia* homeodomain at high resolution (fig. 8.3). Nineteen conformers, the results of nineteen independent distance geometry calculations, are shown in figure 8.3, top, which represents the polypeptide back-

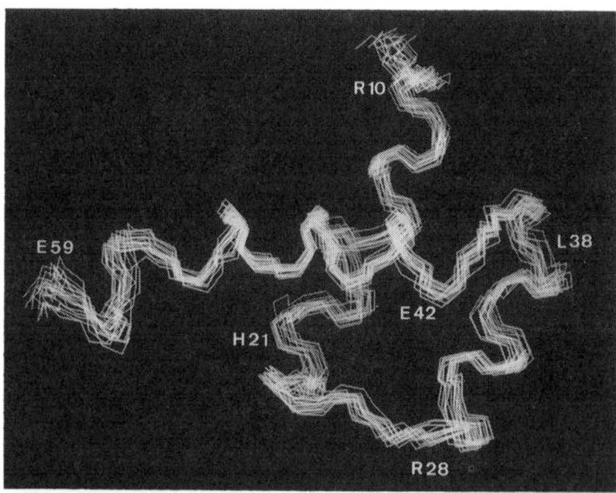

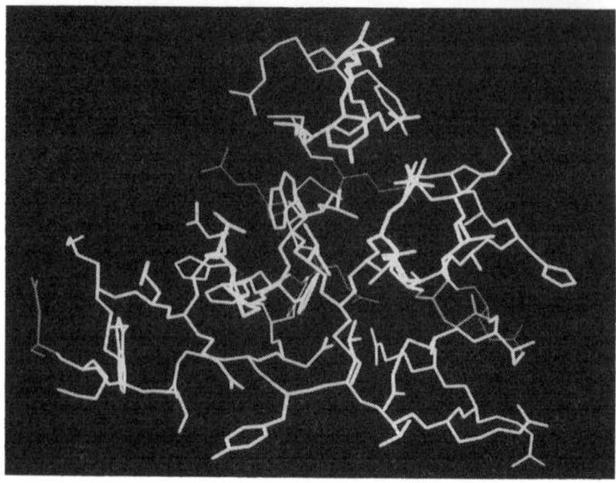

Figure 8.3
Structure of the Antennapedia *homeodomain (as determined by nuclear magnetic resonance spectroscopy).* Top: *Structure of the polypeptide backbone based on 19 independent distance geometry calculations for residues 7–59. Individual amino acids (e.g., R10) are labeled as landmarks.* Bottom: *Structure including all amino acid side chains. After Y.-Q. Qian, M. Billeter, G. Otting, M. Müller, W. J. Gehring, and K. Wüthrich, The structure of the* Antennapedia *homeodomain determined by NMR spectroscopy in solution: Comparison with prokaryotic repressors,* Cell 59 *(1989): 573–80.*

bone. There are three well-defined α-helixes: helix 1 extending from residues 10 to 21, followed by a loop connecting it to helix 2, which runs essentially antiparallel to helix 1. Helix 2 (residues 28 to 38) is connected with helix 3 (residues 42 to 52) by a short turn of three amino acids, forming the well-known helix-turn-helix motif first discovered in various prokaryotic repressors. Helix 3 is extended by a more flexible fourth helix (residues 53 to 59). The amino terminal arm (residues 1 to 6) is flexibly disordered, as are residues 60 to 68 extending beyond the C-terminus of the homeodomain. The three helical regions are folded into a tight globular structure with eleven amino acids, forming a well-defined hydrophobic core. The hydrophilic residues are mostly on the surface of the molecule with the basic side chains oriented toward the DNA (see below) and the acidic amino acids oriented toward the "backside." The MATα2 homeodomain, which shares only 28 percent amino acid sequence identity and carries an insertion of three amino acid residues when compared to *Antennapedia,* has a very similar three-dimensional structure, as Cynthia Wolberger and Sandy Johnson determined in cocrystals of the MATα2 homeodomain with its DNA binding site. The additional tripeptide segment is accommodated at the C-terminus of helix 1 in the loop connecting helixes 1 and 2 without affecting the global structure significantly. Therefore, as John Shepherd predicted, yeast definitively has homeodomains.

A few years later Eric Wieschaus and his collaborator Cornelia Rauskolb found that the *Drosophila* homeodomain encoded by the *extradenticle* (*exd*) gene also has an insertion of three amino acids at the same position as MATα2. When Eric had left my laboratory after completing his graduate studies, he had sworn that he would never clone a gene; apparently, he couldn't avoid it, however, and the first gene his group cloned was *extradenticle* and it had a homeobox! So much for the promises of my graduate students.

A three amino acid insertion has also been found in the human *prl* (=*pbx*) genes at this same position. In going back from yeast to bacteria, hardly any sequence identity between the helix-turn-helix motifs of prokaryotic repressors and those of homeotic genes can be detected, but their three-dimensional structures are readily superimposable (fig. 8.4). The global folds of prokaryotic repressors and their mode of DNA binding differ significantly from that of the homeodomain. Yet there is evidence that the bacterial *Hin* recombinase represents in many respects an intermediate between the prototypical prokaryotic helix-turn-helix proteins and the eukaryotic homeodomain proteins. The *Hin* recombinase of *Salmonella* is a site-specific

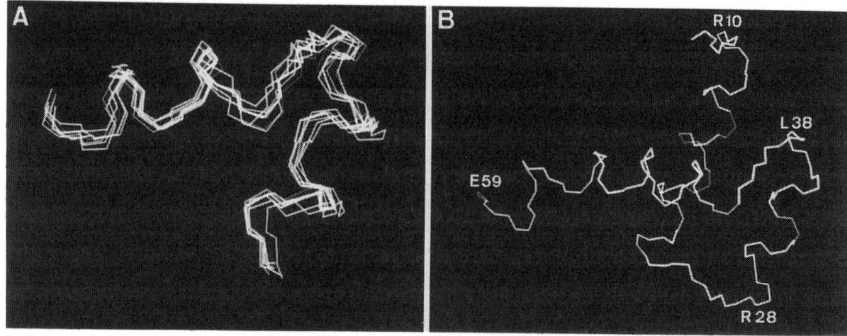

Figure 8.4
Helix-turn-helix motif in Antennapedia *and bacterial repressors. (A) Superposition of the helix-turn-helix motif (backbone) of the* Antennapedia *homeodomain and the corresponding motifs of six bacterial repressor proteins. (B) Complete backbone structure of the* Antennapedia *homeodomain in the same orientation for comparison. Individual amino acids are labeled as landmarks. The helix II extends from R28 to L38, followed by three amino acids in the turn. Helix III/IV extends from E42 to E59. After W. J. Gehring, M. Müller, M. Affolter, A. Percival Smith, M. Billeter, Y.-Q. Qian, G. Otting, and K. Wüthrich, The structure of the homeodomain and its functional implications,* Trends in Genetics *6 (1990): 323–29.*

recombinase that controls the alternate expression of two flagellin genes (encoding flagellar proteins) by reversibly switching the orientation of a promoter. As Markus Affolter discovered, the amino acid sequence of the Hin DNA-binding domain can be aligned with that of the *engrailed* homeodomain and clearly shows some sequence similarity. Recent structural studies of the Hin recombinase-DNA complex have confirmed the similarity in structure and mode of DNA binding. The Hin recombinase thus represents an intermediate combining some of the prokaryotic features of the helix-turn-helix motif with a mode of DNA binding that is reminiscent of that of the homeodomain. It is interesting to note that the Hin recombinase in *Salmonella*, MATα2 in yeast, and the homeodomain proteins of the fungus *Ustilago* are all involved in the genetic control of primitive forms of cellular differentiation, which is similar to the function of homeodomain proteins in higher organisms.

To study the homeodomain-DNA complex by NMR spectroscopy we had to label the protein. In our first experiment we used the isotope nitrogen 15 (^{15}N) for this purpose and analyzed the complex of labeled *Antennapedia* homeodomain with its fourteen base pair DNA binding site.

As Tony Percival-Smith's biochemical studies suggested, the homeodomain-DNA contacts were not confined to the recognition helixes (3 and 4) binding in the major groove of the DNA; the flexibly disordered N-terminal arm also established contacts in the minor groove. To get more detailed structural information, particularly about the amino acid side chains, the protein had to be labeled by carbon 13 (^{13}C). This was an expensive experiment because the ^{13}C label alone costs several thousand dollars even if one uses acetate as a carbon source to grow the bacteria rather than glucose, which is even more costly. This meant that I had to raise special funds for the experiment, and Martin Müller was quite nervous because we couldn't afford to repeat it a second time. And so, instead of growing the bacteria in one big fermentation tank, we distributed the risk of contamination from other microorganisms to several smaller flasks (which is typically Swiss; on a per capita basis Switzerland spends more money on insurance than any other country in the world). All the cultures grew well, and the purification of the ^{13}C-labeled homeodomain, which was done in several batches, also proceeded rather well. Again we carried the labeled homeodomain preparation to Zurich, where the DNA binding site had been synthesized.

The following day, Kurt Wüthrich called me over the phone and told me simply that upon addition of the protein to the DNA, the complex had precipitated and now we had a few thousand dollars lying at the bottom of this small test tube. He then went on to consider what could have gone wrong, and after some discussion we both thought that most likely we did not remove all the salt from our preparation. At that point, we decided simply to dialyze the precipitate against distilled water to remove the leftover salt. I was not in a good mood that night, but he called again the next morning to tell me that the precipitate had dissolved. What a relief!

Now Kurt Wüthrich and his crew could solve the three-dimensional structure of the *Antennapedia* homeodomain-DNA complex at high resolution (pl. 5). A schematic diagram of the complex viewed perpendicular to the recognition helix (3, 4) and along the axis of the recognition helix is presented in figures 8.5A and B, respectively. It can be compared to the corresponding diagrams of the *engrailed* and MATα2 homeodomain-DNA complex structures in 8.5C and D. The latter two were determined by X-ray crystallography in the laboratories of Carl Pabo and Cynthia Wolberger, respectively. The three complexes are very similar both in terms of the overall structure and with respect to the amino acids that make base-specific contacts to the DNA. One of the conformers used to describe the NMR solution

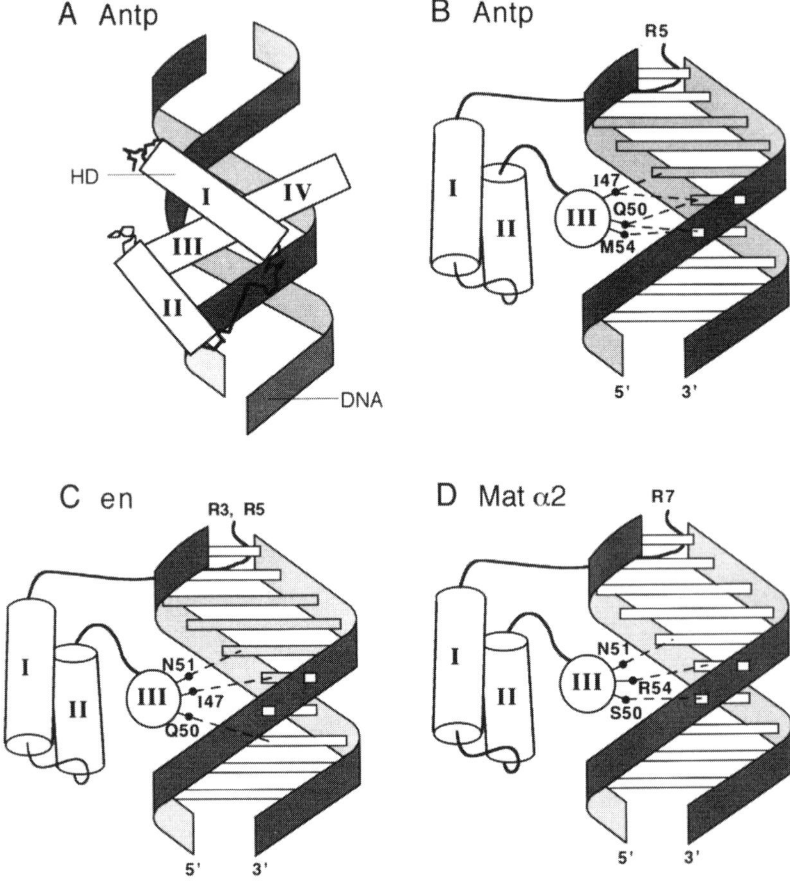

Figure 8.5
Schematic comparison of the homeodomain-DNA complexes of Antennapedia, engrailed, and Matα2. *(A) View perpendicular to the axis of the recognition helix (III/IV) located in the major groove of the DNA. The α-helices (I–IV) are indicated as white cylinders, the homeodomain backbone by a black line, and the DNA backbones by gray ribbons. (B–D) View along the axis of the recognition helix (III). Amino acid residues that establish direct contact to specific bases in the DNA are indicated. The N-terminal arm reaches into the minor groove (with residues R5, R3 and R5, and R7, respectively). The contacts between specific residues in the recognition helix and bases in the major groove are indicated by dotted lines. The base pairs are represented by bars. The base pairs of the ATTA core motif are indicated in gray. After W. J. Gehring, Y.-Q. Qian, M. Billeter, K. Furukubo-Tokunaga, A. F. Schier, D. Resendez-Perez, G. Otting, and K. Wüthrich, Homeodomain-DNA recognition,* Cell *78 (1994): 211–23.*

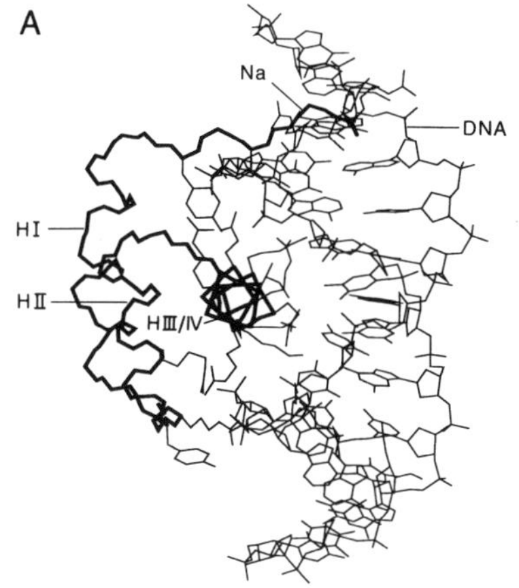

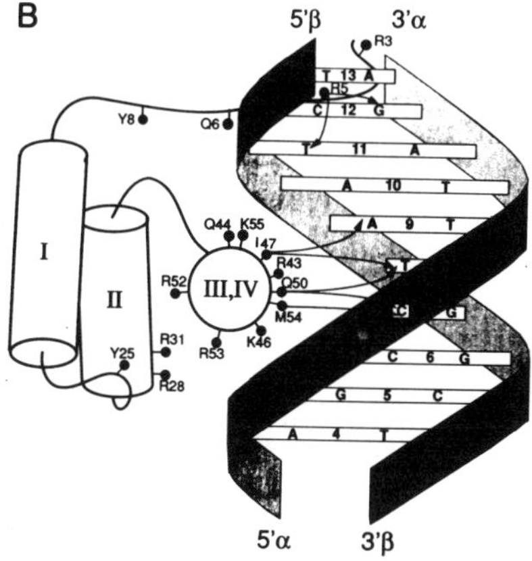

Figure 8.6
The Antennapedia *homeodomain-DNA complex. (A) Structure of complex showing only those amino acid side chains that contact the DNA. The DNA double helix is on the right. The backbone of the homeodomain is represented by a dark line. The N-terminal arm (Na) reaches into the minor groove of the DNA. Side chains from the recognition helix (HIII/IV) contact the bases and the backbone of the DNA in the major groove. Courtesy of K. Wüthrich. (B) Schematic drawing of the complex. The α-helixes of the homeodomain are indicated as cylinders, the backbone as a blackline. The DNA backbones are represented as ribbons (α and β) with their 5′ and 3′ ends indicated, the DNA base pairs as bars labeled 4 to 13. Amino acid residues contacting the DNA are indicated by black circles, their base-specific contacts by arrows. From W. J. Gehring, Y.-Q. Qian, M. Billeter, K. Furukubo-Tokunaga, A. F. Schier, D. Resendez-Perez, G. Otting, and K. Wüthrich, Homeodomain-DNA recognition,* Cell *78 (1994): 211–23.*

structure of the *Antennapedia* homeodomain-DNA complex is shown in figure 8.6. Only those amino acid side chains that contact the DNA are indicated. The major areas of contact are between amino acid side chains in the recognition helix and the major groove of the DNA, the flexible N-terminal arm that contacts the minor groove, and numerous additional contacts to the backbone of the DNA.

The availability of these detailed structural data allowed us to consider the question of the binding specificity. How did the homeodomain recognize these fourteen base pairs? The core motif ATTA (or TAAT on the opposite DNA strand) was recognized by all the homeodomains that were studied, but this was obviously insufficient to account for the binding specificity in vivo. There is a genetic experiment that allows you to find out which amino acid of a DNA binding protein interacts with which base in the DNA. This experiment was carried out previously with the CAP (catabolite repression protein) of *E. coli,* and to my mind, it demonstrates the power of bacterial genetics most impressively. The experiment carried out by Richard Ebright, Pascale Cossart, Brigitte Giquel-Sancey, and Jon Beckwith consisted of first mutating one base pair in the DNA binding site, the so-called operator, and then looking for a compensating amino acid exchange in the CAP protein by selecting among hundreds of millions of mutagenized bacteria (more than 10^8) for those few whose CAP protein would recognize the altered DNA of the operator binding site. They found very few such mutants, but all of them affected a single amino acid of the CAP protein specifically. This is called a second site suppression experiment. The effect of a mutation in one gene, the operator, is suppressed by a mutation in another gene, in CAP. When the structure of the CAP-operator-DNA complex was determined several years later, it was found that the mutated base pair is indeed in contact with the mutated amino acid. This was one of my dream experiments.

But although you can select for a rare mutation among 10^8 bacteria, breeding 10^8 flies is a hopeless undertaking, and we had to find a different solution. The amino acid sequence of the recognition helix in the various homeobox genes displays relatively little variation; only at position 50 can one find significant differences. The homeodomains of *Antennapedia, fushi tarazu,* and the entire *Antennapedia* gene family, for example, have a glutamine at position 50, whereas *bicoid* has a lysine residue at this position. On the basis of Mark Ptashne's work on the Lambda repressor, which is elegantly described in his little book *A Genetic Switch,* one would have expected the amino acid at position 43 (the second amino acid of the recognition helix) to be of crucial importance, but the homeodomain differs in its mode of binding from that of the Lambda repressor; its recognition helix is positioned differently in the major groove, and it also binds as a monomer rather than as a dimer. Roger Brent, a former student of Mark Ptashne, and Claude Desplan had obtained evidence independently that position 50 of the homeodomain was important for DNA binding specificity. But it was not known which base in the binding site was contacted by the amino acid at position 50. By comparing the binding sites of *bicoid,* which has a lysine at position 50, and that of *fushi tarazu,* with a glutamine at this position, Markus Affolter in my laboratory was able to predict the two bases that are concerned: the *bicoid* consensus binding site determined by Wolfgang Driever is GGGATTAGA, whereas the consensus binding site for *fushi tarazu* and *Antennapedia* as determined by Markus Affolter is GCCATTAGA. The only differences are thus found at positions 6 and 7, where *bicoid* has GG and *fushi tarazu* has CC. This prediction was tested in the context of the *fushi tarazu* gene, which subsequently allowed us to do the second site suppression experiment in vivo. For this purpose we first constructed a *fushi tarazu* mutant homeodomain with lysine (K) at position 50 (*ftz*50K) rather than glutamine (*ftz*50Q). The measurements of the binding constants for the four possible binding sites with the sequences CC, CG, GC, and GG at positions 6 and 7 showed that *ftz*50Q has the highest affinity fo CC, whereas *ftz*50K preferentially binds to GG. This was consistent with the NMR data, which showed that glutamine 50 indeed contacts C7 and C6 in the binding site (see below).

On the basis of this structural and biochemical information, we were now in a position to design the second site suppression experiment rather than having to rely on the isolation of a rare mutation. My former graduate student Alexander Schier carried out the experiments. For this purpose we used the autoregulatory enhancer element of the *fushi tarazu* gene. This 430 base pair minimal autoregulatory element contains one high-affinity and five medium-affinity in vitro binding sites for the fushi tarazu protein. In transgenic embryos this element, when fused to a β-galactosidase reporter gene, directs reporter gene expression in a *fushi tarazu*–like pattern of seven stripes, which depends on the expression of the normal fushi tarazu protein in those embryos. Deletion analysis of the in vitro binding sites indicated that these multiple binding sites are functionally redundant. However, when three or more of these binding sites are either deleted or inactivated by point mutations, reporter gene expression is essentially abolished. For the purpose

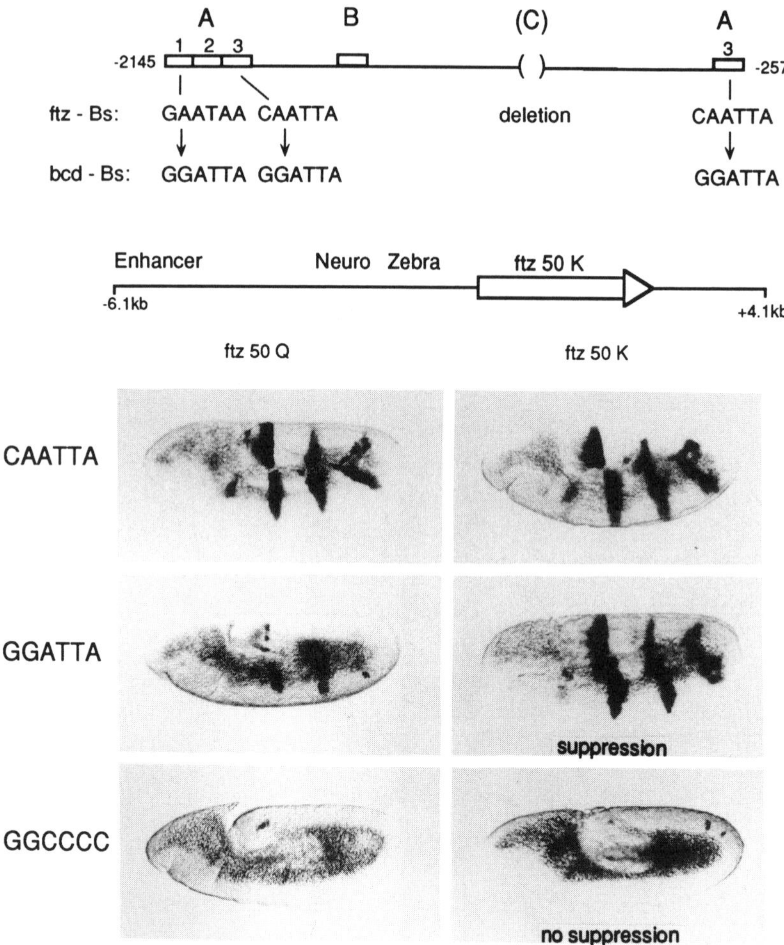

Figure 8.7
Evidence for direct interaction of the fushi tarazu protein with the DNA binding sites. Top: *Autoregulatory element, a DNA segment containing four putative binding sites* (ftz-*Bs), one of which has been deleted (C). The three remaining sites were mutated to bicoid binding sites* (bcd-*Bs) of the sequence GGATTA.* Middle: fushi tarazu *gene in which amino acid 50, glutamine* (ftz-*50Q), is replaced by lysine* (ftz-*50K).* Bottom: *Expression pattern of the reporter gene (β-galactosidase) fused to an autoregulatory element with either* fushi tarazu *binding sites (CAATTA),* bicoid (bcd) *binding sites (GGATTA), or randomly mutated sites (GGCCCC) in combination with either* ftz-*50Q or* ftz-*50K.* ftz-*50Q induces the normal seven stripes with the* ftz *binding sites but activates only weakly through* bcd *binding sites, whereas* ftz-*50K induces seven stripes with both* ftz *and* bcd *binding sites. Finally, neither protein induces seven stripes with the random binding sites. These findings indicate that the effects of mutating the three* ftz *binding sites to three* bcd *binding sites can be suppressed by a compensatory change in the protein from glutamine (50Q) to lysine (50K).*

of the second site suppression experiment, three of the in vitro binding sites were point mutated to GGATTA, and one was deleted entirely. Transgenic embryos with a normal fushi tarazu protein (ftz50Q) carrying these mutated binding sites show hardly any reporter gene expression (fig. 8.7). Introducing a transgene encoding a fushi tarazu (50K) protein, however, restored the expression pattern of the seven stripes. This experiment proved two important points: first, the interaction between the fushi tarazu protein and its autoregulatory element in the living embryo is direct, and second, the modes of DNA binding in vivo and in vitro must be very similar. As a biologist, I found this confirmation of the NMR data by in vivo experiments very reassuring.

The NMR data revealed several very interesting dynamic aspects that I would like to mention here because they have a considerable influence on our way of thinking about protein-DNA interactions. The first concerns the internal mobility of the homeodomain. The detailed analysis of the contacts between glutamine 50 and the DNA indicates that a single amino acid can contact two or even three bases in the different conformers. Such contacts are observed in most conformers from glutamine 50 to C7 and T8, but an NOE (nuclear Overhouser effect) was also detected to C6. Furthermore, the NMR studies revealed the presence of hydration water molecules at the interface between homeodomain and DNA. These water molecules might mediate specific hydrogen bonds between glutamine 50 and asparagine 51 and polar groups on the DNA whenever the interatomic distances are too large to enable the formation of direct hydrogen bonds. If we adopt the hypothesis that these structural variations represent conformational changes in time, we get the impression that the amino acid side chains scan the DNA surface rather than being locked in a fixed position.

The DNA binding and the functional specificity of homeodomain proteins reside not only in the recognition helix but to a large extent also in the N-terminal arm of the homeodomain. In the free protein, the N-terminal arm is flexibly disordered, and upon binding to DNA it establishes DNA contacts in the minor groove. Genetic experiments show that the N-terminal arm of the homeodomain contributes significantly to the functional specificity of those homeoproteins like Antennapedia and Sex combs reduced that have identical recognition helixes and differ significantly in their function. The homeodomains of these two proteins differ at five positions only, positions 1, 4, 6, and 7 in the N-terminal arm and position 60 at the C-terminus. With experiments in which parts of the coding regions were

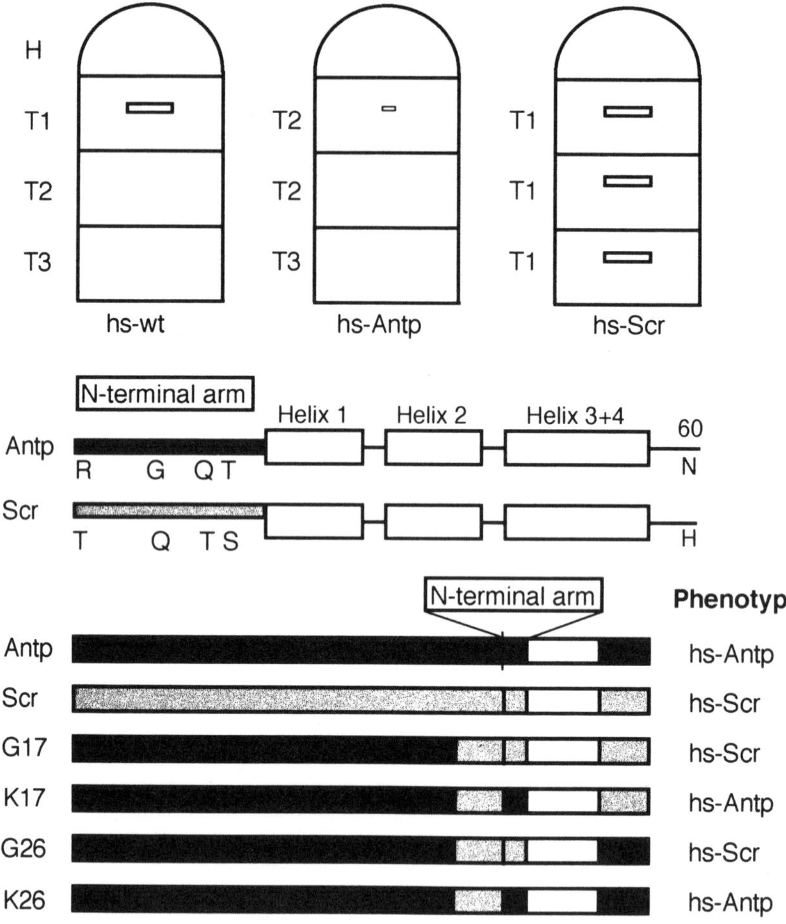

Figure 8.8
Functional specificity in the N-terminal arms of the homeodomain in Antennapedia *and* Sex combs reduced. Top: *Schematic representation of the anterior segments of the larva (H head, T1–3 thoracic segments). The rectangles inside the segments indicate the beard-like structure characteristic for T1. The heat-shocked wild-type larva (hs-wt) shows the normal pattern, the heat-induced* Antennapedia *(hs-*Antp*) shows a transformation of T1 to T2 (reduced beard), and the heat-induced* Sex combs reduced *(*Scr*) transforms both T2 and T3 to T1 (with beards).* Middle: *Amino acid differences between* Antp *and* Scr *in the N-terminal arm.* Bottom: *Phenotypes of the various constructs. The phenotype (whether hs-*Antp *or hs-*Scr*) always corresponds to the N-terminal arm. Courtesy of K. Furukubo-Tokunaga.*

exchanged between these two genes, Greg Gibson clearly showed that the functional specificity of the two proteins resides in or very near the homeodomain. In a heat-shock assay with transgenic flies producing different chimeric Antennapedia–Sex combs reduced proteins, additional first thoracic (T1) segments are induced if the homeodomain originates from *Sex combs reduced,* and additional T2 segments are induced whenever the chimeric protein contains an *Antennapedia* homeodomain (fig. 8.8). By substituting the four amino acids that are specific for *Sex combs reduced* with those of *Antennapedia,* Katsuo Furukubo-Tokunaga was able to show that the functional specificity resides in those four amino acid residues only (see fig. 8.7). The reciprocal experiment was carried out in Matthew Scott's laboratory and led to the same conclusion. Because the NMR data show that of these four amino acids only glutamine 6 contacts the DNA, we think not that the functional specificity is determined exclusively by the DNA binding specificity but rather that additional protein-protein interactions are involved.

On the basis of the combinatorial model for gene regulation, one homeodomain protein seems to bind not alone to DNA but in conjunction with other transcription factors and to form multimeric complexes either in solution or upon binding to DNA. Such combinatorial interactions have been clearly demonstrated for the yeast MATα2 protein. MATα2 can either associate with the MCM1 protein and form a hetero-tetramer, which recognizes a set of operator sites associated with genes that are specifically expressed in cells of mating type a, or alternatively it forms a heterodimer with the MATa1 homeodomain protein and binds to a set of haploid-specific genes. The same protein in a different context may therefore serve a different function. The interaction between MATα2 and MCM1 is mediated by an apparently unstructured region of the MATα2 protein that may correspond to the flexible linker arm, the N-terminal of the *Antennapedia* homeodomain, which includes the YPWM motif that is conserved in evolution from *Drosophila* to humans. This linker arm may therefore also be involved in protein-protein interactions in higher organisms. Further, protein-protein interactions have been shown to occur between the homeodomain protein of *extradenticle* on one hand and *Ultrabithorax* and *Antennapedia* on the other, but we are still just beginning to unravel these interactions.

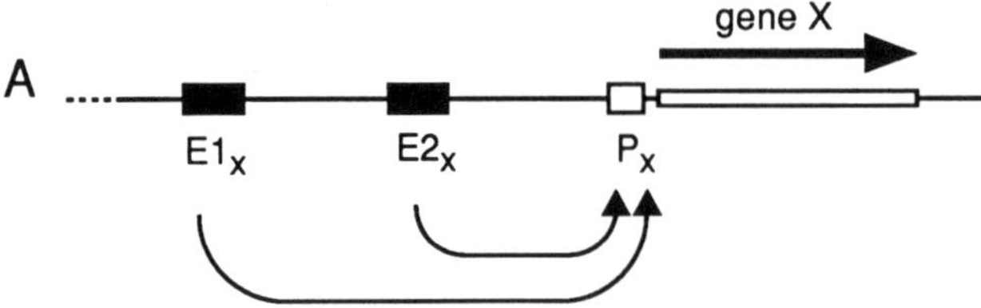

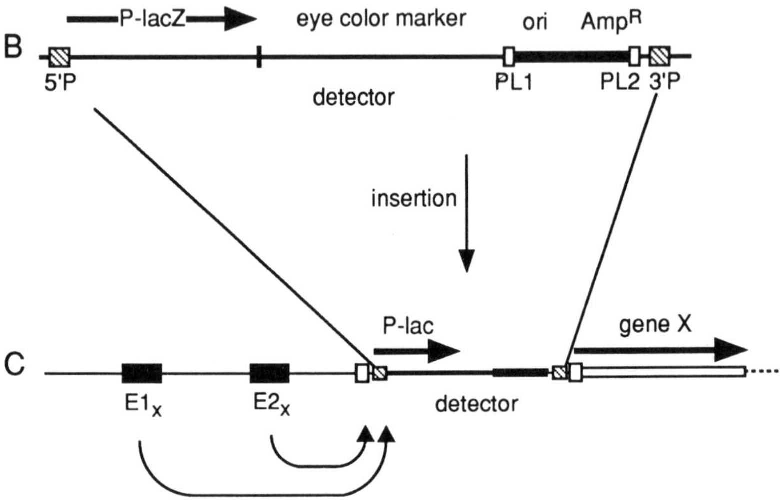

Figure 9.1
*Principles of enhancer detection. (A) Enhancers ($E1_x$ and $E2_x$) are regulatory elements on the DNA to which specific activator and repressor proteins bind and exert a regulatory effect on the nearest promoter (Px), thereby regulating gene expression at some distance either upstream or downstream of the gene (X). (B) Enhancer detector plasmid. P-transposons with 5' and 3' flanking sequences (5' P and 3' P), a ubiquitously expressed minimal promoter fused the reporter gene β-galactosidase (P-*lacZ*), an eye color marker, and a bacterial vector (with two polylinkers PL1+2, an origin of replication* ori, *and an Ampicillin-resistance gene* Amp^R*). (C) Insertion of the detector plasmid in the vicinity of a gene (X). The enhancers $E1_x$ and $E2_x$ act upon the minimal promoter and drive P-*lac *in a pattern similar or identical to gene X.*

~

9
Enhancer Trapping and Hunting for Target Genes

WHEN LOOKING at the beautiful pattern of seven stripes that were generated by the fusion of the *fushi tarazu* enhancer to the β-galactosidase reporter gene it occurred to me that it should be possible to turn things around and to use the reporter gene to detect unknown enhancers. This rather vague idea began to crystallize when Cahir O'Kane, an Irishman who had trained in bacterial genetics, joined my laboratory. The year was 1986 and operon fusion experiments were "en vogue"; he wanted to apply this technique to *Drosophila* genetics where it had never been used before. These experiments consist of inserting a promoterless reporter gene at random into the bacterial genome in order to find operons. An operon is a small group of genes that share a common promoter and are transcribed into a single messenger RNA, which comprises all the genes of the operon. Upon insertion of the reporter gene into an operon, it is transcribed under the control of the promoter in the same way as the other genes in the respective operon.

There is an essential difference in the genome organization of bacteria and higher organisms, however; bacterial genes are organized into operons, whereas *Drosophila* genes are regulated individually. In higher organisms each gene is transcribed from its own promoter as a single unit, and a particular DNA sequence, the TATA box, has to be close to the start site of transcription. Further, the bacterial genomes are very compact with tiny intergenic regions, whereas the genes in higher organisms are more widely spaced. This meant that we could not use a promoterless reporter gene in *Drosophila* for enhancer detection. O'Kane and I discussed this problem at length, although I sometimes had considerable difficulties in understanding his Irish accent. After several hours of discussion we drew up different schemes, some relatively simple, some more sophisticated, which he was going to try. Meanwhile, in Zurich Walter Schaffner had constructed what he called an "enhancer trap" for the polyoma virus. This virus contains a regulatory DNA segment that was one of the first enhancers to be discovered. It enhances gene expression in cis, which means if located on the same DNA molecule as the target gene, at a relatively large distance (in terms of base pairs), and in both orientations—that is, from upstream or downstream of the target gene on the circular viral DNA molecule. By removing the enhancer Schaffner generated a virus that replicated very poorly and could serve as an enhancer trap. By cloning random pieces of DNA into this defective virus he could select for DNA segments with enhancer activity. This is not what we had in mind for *Drosophila*. We wanted to detect enhancers and other gene regulatory elements in the *Drosophila* genome, and so we called the procedure enhancer detection. When the method was established and American laboratories adopted it, however, they called it simply enhancer trapping.

As is usual in life, the simplest of the schemes that O'Kane and I designed worked best. It makes use of the P-transposon both for random integration into the genome and for providing a promoter to the reporter gene. It was known from Gerry Rubin's work that the promoter of the transposase gene is located close to the 5′ end of the P-transposon and that it is active in all cells of *Drosophila*. Because the transposase promoter is relatively weak, it was ideally suited as a sensor for gene regulatory elements. O'Kane replaced the transposase gene by the β-galactosidase reporter and included the *rosy* (*ry*) eye-color gene as a marker for detecting transposition. The principle of enhancer detection basically involves the random integration of the detector transposon by germline transformation into many different sites in the genome (fig. 9.1). When the detector transposon inserts near a genomic enhancer, the reporter gene is expressed under the influence of the enhancer, which can be detected by staining for β-galactosidase activity. When we designed the experiment, O'Kane and I thought that in the best case a few percent of the insertions might detect an enhancer; if, however, the probability was lower and we had to examine more than a thousand insertions to find a single enhancer, the procedure would be useless. Our colleagues' estimates were in the same order of magnitude. Then Yash Hiromi injected the detector transposon and O'Kane recovered some sixty transformants. Our expectations had been surpassed by far; more than half of the insertions showed developmental regulation of reporter gene expression in the embryo. It turns out that the *Drosophila* genome is loaded with genomic regulatory elements driving reporter gene expression, and the procedure opened a little gold mine.

Clive Wilson subsequently improved the detector plasmid by adding bacterial vector sequences that allowed the *Drosophila* DNA sequences that flanked the insertion site to be easily cloned, and Hugo Bellen and Ueli Grossniklaus introduced the jump-starter procedure developed in Allen Spradling's laboratory that allows new insertions (transposants, as we called

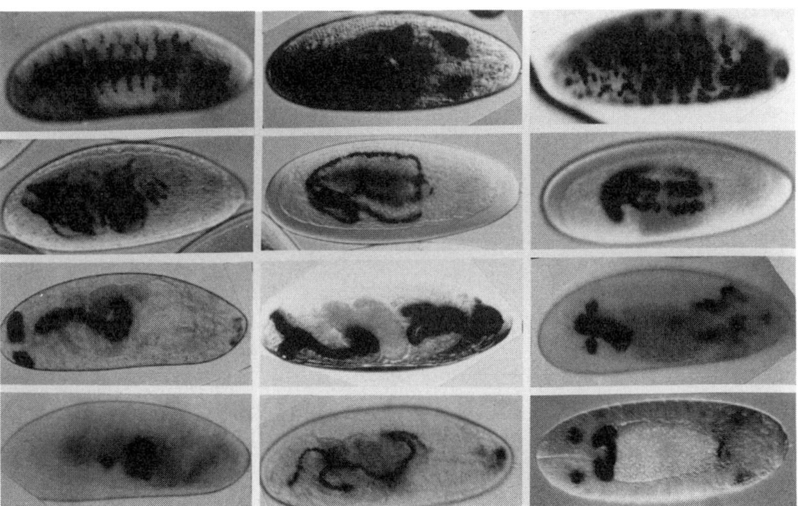

Figure 9.2
Examples of expression patterns of the β-galactosidase reporter gene in embryos of different enhancer detector lines. After H. J. Bellen, C. J. O'Kane, C. Wilson, U. Grossniklaus, R. Kurth Pearson, and W. J. Gehring, P-element-mediated enhancer detection: A versatile method to study development in Drosophila, Genes and Development *3 (1989): 1288–1310.*

them) to be isolated by simple genetic crosses to a stable transposase, producing stock isolated by Bill Engels. Activating the transposase enzyme mobilizes the resident transposons. Using these stocks, some 550 transposants were isolated and characterized by a congenial crew called the "jump-start team" consisting of Hugo Bellen, Clive Wilson, Ueli Grossniklaus, Cahir O'Kane, and Becky Pearson. Eighty-five percent of the transposants showed β-galactosidase staining and identified genomic regulatory elements. Such a surprising efficiency might lead one to suppose that many of these elements were cryptic and do not normally regulate *Drosophila* genes. Many of our colleagues remained highly skeptical, and Gary Struhl proclaimed, "These are not genes." Yet evidence now suggests that in most transposants the regulatory sequences revealed by the enhancer detector do normally control *Drosophila* genes. This procedure can thus be used for detecting genes with an interesting pattern of expression. With the bacterial vector sequences such genes can be readily cloned, and deletion mutants can be obtained by "jumping out" of the detector transposon, which is often imprecise and deletes some of the flanking sequences. Bier and coworkers isolated no fewer than 3,500 transposants, confirming our results. Since then, enhancer detection has become a valuable tool in *Drosophila* genetics, and a modified version of the procedure has been applied to mouse and plant genetics.

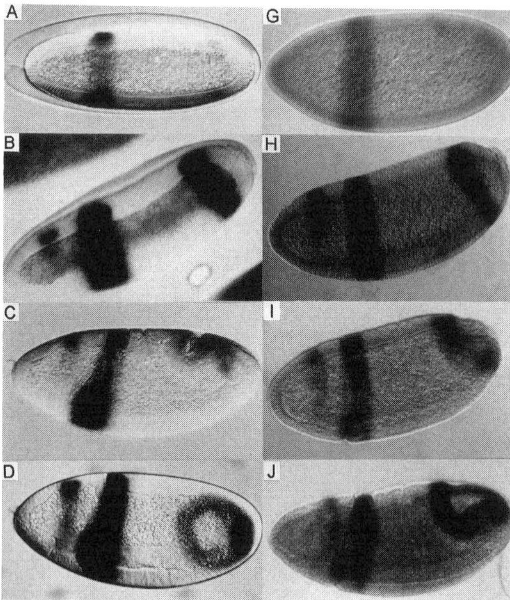

Figure 9.3
Comparison of the expression patterns of the enhancer trap line (staining for β-galactosidase in A–D) with the expression pattern of the spalt *gene (in situ hybridization in G–J) at the blastoderm and gastrula stages. After J. Wagner-Bernholz, Identification of target genes of the homeotic gene* Antennapedia *by enhancer detection (Ph.D. Thesis, University of Basel, 1991), fig. 11.*

A spectrum of the expression patterns in different transposant lines is shown in figure 9.2 for embryos and in plate 6 for imaginal discs. The detected gene regulatory elements are generally enhancers that can be used for targeted gene expression, a crucial genetic tool explained later in this book. Another critical application is gene isolation. You can isolate genes that are associated with the detected enhancer by cloning the DNA sequences flanking the point of insertion and ask whether the pattern of expression of the cloned gene is the same as that of the reporter gene in the transposant line. The answer is generally yes, unless the gene has a complicated structural organization, with, for example, two promoters and different enhancers. In this case, the reporter may reflect only part of the expression pattern. In the *spalt* (*sal*) gene, for example, the pattern of reporter gene expression coincides with the pattern of *spalt* as determined by in situ hybridization to *spalt* messenger RNA (fig. 9.3). Enhancer detection is thus a convenient method for gene isolation based on the expression pattern. It has two main advantages over standard mutagenesis: the respective gene can easily be cloned, and it allows the isolation of redundant genes. Functionally redundant genes are not detected by chemical mutagenesis because knocking out one gene alone has little effect, and double mutants are difficult to obtain.

Enhancer detection also allows the isolation of genes that are not required for viability and whose phenotype may not be easily detectable in loss-of-function mutations. Among the first transposant lines isolated was one that showed a *fushi tarazu*–like pattern of stripes. This was a complete surprise, because we originally thought that such insertions would be extremely rare. Ueli Grossniklaus and Becky Pearson later cloned the gene, and it turned out to be *sloppy paired*, an interesting case of gene redundancy. As Grossniklaus and Ken Cadigan showed, *sloppy paired* is a pair of genes (*slp* 1 and *slp* 2) that arose by gene duplication and still shares most of the gene regulatory elements. The two genes have essentially the same spatial pattern of expression and differ only in the timing of their expression. *sloppy paired* was detected by chemical mutagenesis, but only because the two genes are very closely linked so that a mutation affecting both genes could be isolated. The other example I mention here is fasciclin III, a gene involved in the formation of bundles of nerve fibers and nerve path finding. As shown by Corey Goodman, inactivation of the fasciclin III gene does not result in lethality because of its functional redundancy with some of the other fasciclin genes. However, fasciclin III can be detected by enhancer trapping.

Enhancer detection also allows the identification of genes required at multiple stages of development and genes with subtle phenotypes that may easily be missed in standard genetic screens.

A further application of enhancer detection concerns the identification of target genes. So far we have mostly considered the genetic control exerted by homeotic genes at the transcriptional level high up in the regulatory cascade. At early syncytial stages, before the blastoderm stage when cell membranes form, transcription factor proteins can function as signaling molecules to provide positional information. They contain nuclear targeting sequences and can penetrate the nuclei through the pores in the nuclear envelope (fig. 9.4). Once cell membranes are formed, different signaling mechanisms have to be used. For this reason some researchers have suggested that *Drosophila* is a special case and that the results obtained in *Drosophila* may not be applicable to vertebrate embryos, which have a different mode of cleavage. Yet I do not think that this is such a fundamental difference, because cleavage in zebra fish or chickens is also incomplete at early stages and the blastomeres are still connected to each other via a large yolk cell (or yolk sac). In addition, Yoshiaki Suzuki and his collaborators have found that *caudal* transcripts form a gradient in the silk moth embryo much as in *Drosophila* even though the silk moth embryo is cellularized, not syncytial as in *Drosophila*. In any case, once cell membranes are formed, a signal has to be sent from the nucleus to the cell surface and communicated to the corresponding receptor on the surface of the target cell. The activated receptor triggers a signal cas-

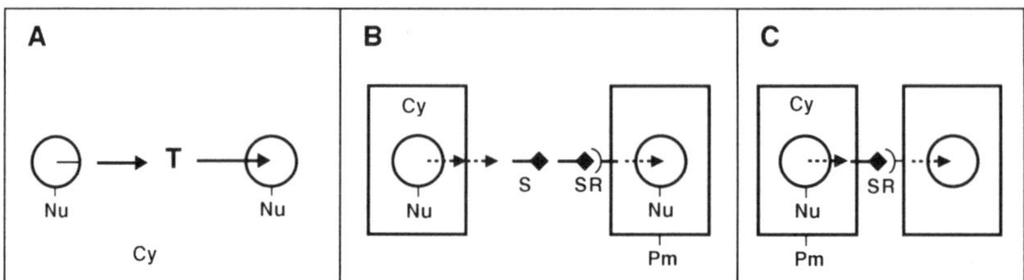

Figure 9.4
Signaling between nuclei and cells: (A) in syncytia (in the absence of cell membranes), the nuclei can express a transcription factor (T), which can penetrate into another nucleus and regulate gene activity directly; (B) signaling via a diffusible signaling molecule emanating from a signaling cell and interacting with a membrane-bound receptor on the target cell and initiating a signal transduction cascade; (C) signaling via a membrane-bound signaling molecule for short-range interactions.

cade leading from the cell surface to the nucleus of the target cell, where the signal elicits a response at the level of gene regulation. The signal between the two cells may be a diffusible ligand that is either secreted from the signaling cell or cleaved of its surface, and it diffuses to the receptors on the receiving cell (fig. 9.4). Alternatively, the ligand may never leave the surface of the signaling cell and the response may be triggered only by direct cell-to-cell contact, the two cells "kissing." The former mechanism accounts for a long-range affect of the ligand, whereas the latter mechanism is more suited for the generation of fine-grained patterns, as in eye morphogenesis.

To elucidate the genetic circuits underlying this signaling cascade, the target genes of master control genes have to be identified and isolated. Several methods have been developed for the isolation of target genes regulated by transcription factors. The first target gene of the homeotic *Ultrabithorax* gene was isolated by Rob White and his collaborators by cloning DNA fragments to which the Ultrabithorax protein was bound in chromatin preparations. The *connectin* gene that was cloned in this way is discussed in the following chapter. A common strategy for finding target genes involves the isolation of mutants that either enhance or suppress the phenotype of a homeotic mutation. The genes affected by such mutations are likely to be involved in the same developmental pathway. In my laboratory, we adapted the enhancer detection method to the identification of *Antennapedia* target genes. Juliane Wagner-Bernholz and Susanne Flister first screened our collection of 550 enhancer detector lines for genes that are expressed in the leg or antenna imaginal discs. Then such lines were crossed with *Antennapedia*^{73b} or *Nasobemia*—dominant gain-of-function mutants. If the gene corresponding to the enhancer detector is regulated by *Antennapedia*, one would expect an alteration of the expression pattern in the antennal disc, where the Antennapedia protein is ectopically expressed. Two classes of detector lines were identified, one being activated, the other being repressed by the ectopic expression of the Antennapedia protein. Among the first class of lines, the *klecks* gene is expressed in two characteristic crescents in the leg imaginal discs, and no expression can be detected in the eye-antennal disc. However, upon induction of *Antennapedia* in all imaginal discs, the two crescents appear in the antennal disc (fig. 9.5). One of the lines in the second class that is repressed by *Antennapedia* was analyzed further by Juliane Wagner-Bernholz and shown to represent the *spalt* gene. *spalt* is normally expressed in a ring-shaped pattern in the antennal disc, whereas no expression is observed in the leg discs. In *Antennapedia* gain-of-function mutants the expression in the antennal ring is repressed (fig. 9.5). To test whether *Antennapedia* loss-of-function mutants have the opposite effect, Wagner-Bernholz generated clones of homozygous *Antennapedia*⁻ cells. As expected, such clones exhibit leg-to-antenna transformations and *spalt* is also derepressed in the second leg discs. The finding that *spalt* is repressed in antennal discs by gain-of-function mutants and conversely derepressed in second leg discs by loss-of-function mutants indicates that *spalt* is a down-

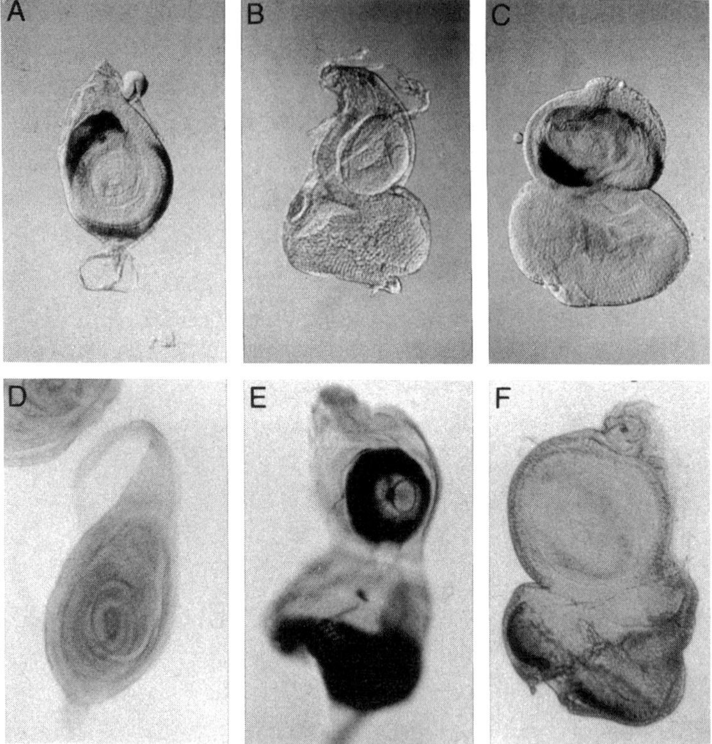

Figure 9.5
Identification of target genes that are either activated or repressed by Antennapedia *with enhancer trap lines. A–C* klecks, *D–F* spalt: *(A) normal expression of* klecks *in the leg disc; (B) lack of* klecks *expression in antenna disc; (C) induction of* klecks *in the antenna disc by ubiquitous expression of* Antennapedia *(hsp-*Antp *after heat induction)—the pattern clearly resembles that in a leg disc; (D) lack of expression of* spalt *in the leg disc; (E) normal expression of* spalt *in a ring-shaped pattern in the antenna disc; (F) repression of* spalt *expression by ubiquitous expression of* Antennapedia *(hsp-*Antp *after heat induction). A–C, Courtesy of S. Flister; D–F, After J. Wagner-Bernholz, C. Wilson, G. Gibson, R. Schuh, and W. J. Gehring, Identification of target genes of the homeotic gene* Antennapedia *by enhancer detection,* Genes and Development 5 *(1991): 2467–80.*

stream target gene that is negatively regulated by *Antennapedia.* The time course of the interaction suggests (but does not prove) a direct interaction. The *spalt* gene was molecularly characterized in collaboration with Reinhard Schuh and Herbert Jäckle, and turned out to encode a zinc finger protein, as have most of the genes Jäckle has characterized. Even though it is downstream of *Antennapedia,* therefore, the *spalt* gene is likely to encode a transcription factor.

Juan Botas has definitively identified a homeotic target gene at a lower level of the regulatory cascade, and this is a gene called *decapentaplegic* (*dpp*). Responsible for this unwieldy name is Bill Gelbart, who has concentrated on the analysis of this gene for many years. *decapentaplegic* encodes a member of the transforming growth factor β (TGFβ) family of proteins that are involved in cell signaling as ligands binding to the corresponding receptors of their target cells. Both *Ultrabithorax* and *decapentaplegic* are expressed in the same spatial domain of the visceral mesoderm of the gut, adjacent to that of the homeotic gene *abdominal-A,* which is expressed more posteriorly. Loss-of-function *Ultrabithorax* mutants lead to a strongly reduced level of *decapentaplegic* expression, whereas *abdominal-A* loss-of-function mutants lead to the ectopic expression of *decapentaplegic* more posteriorly. This suggests that *Ultrabithorax* activates *decapentaplegic,* whereas *abdominal-A* represses it.

To find out whether *Ultrabithorax* directly activates *decapentaplegic* expression, Botas and co-workers used the same strategy that my laboratory used to demonstrate a direct interaction between the fushi tarazu protein and its autoregulatory enhancer. They identified a visceral mesoderm enhancer in the *decapentaplegic* gene, then fused it to the β-galactosidase reporter gene, and determined the in vitro binding sites for *Ultrabithorax.* After showing that these binding sites are required for in vivo enhancer function, they mutated these binding *Ultrabithorax* sites to *bicoid*-type binding sites, which renders them nonfunctional. By introducing a glutamine 50 to lysine mutation into the *Ultrabithorax* gene, they suppressed the mutations in the binding sites and restored the expression pattern of *decapentaplegic.* These experiments demonstrate that the Ultrabithorax protein directly regulates *decapentaplegic* expression.

Because the homeotic master control genes regulate probably hundreds of target genes, many more of these genes will need to be identified before we can see even the outlines of the genetic control circuits that steer morphogenesis.

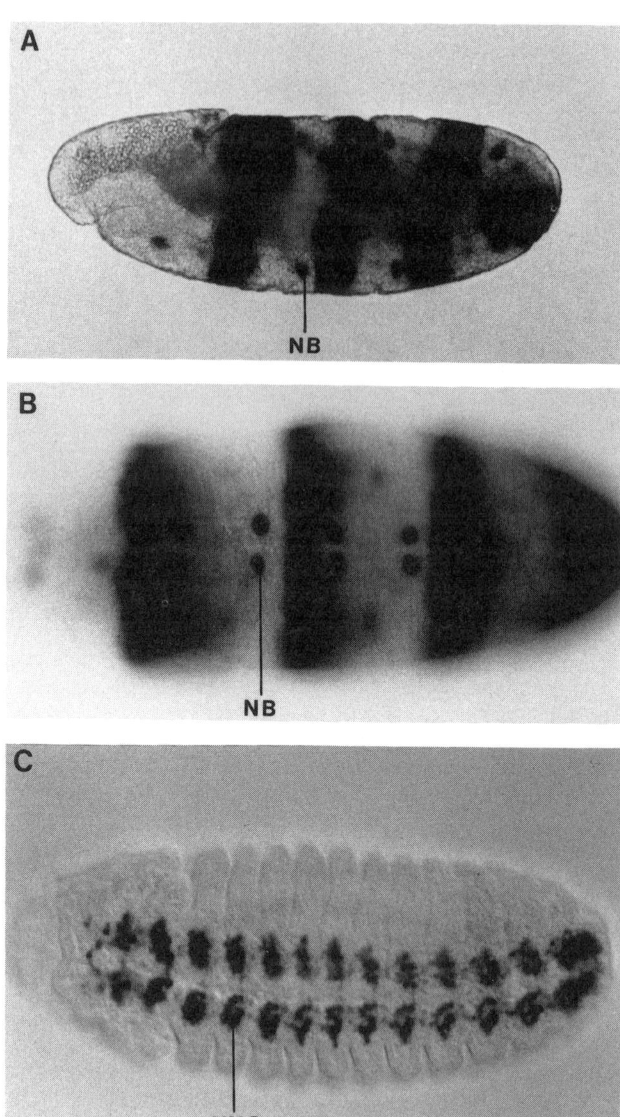

Figure 10.1
Expression of fushi tarazu *in the ventral nervous system: (A) specific pairs of neuronal precursor cells (NB) expressing* fushi tarazu, *lateral view; (B) ventral view; (C) expression pattern in the ventral nervous system (VNS) at later embryonic stages. Courtesy of Y. Hiromi.*

~

10
The Wiring of the Nervous System

How the human brain functions is perhaps the most challenging of all scientific questions. The simple number of neurons (nerve cells) and number of their connections in the human brain are so enormous that the wiring of these cells, or hardware as it's called today, is incredibly complex. If, in addition, one considers the flexibility of the system—the fact that neurons can be disconnected and reconnected with other neurons—then the puzzle becomes almost impossible to solve. The scientific reductionist approach is to tackle the problem first in simpler nervous systems that are amenable to structural and functional analysis. One way is to choose an animal with fewer and giant neurons, such as the sea-hare *Aplysia,* a big marine snail, in which you can do biochemistry and physiology on a single neuron. Another is to choose an organism that is amenable to molecular genetic analysis, such as *Drosophila.*

I, however, had not the slightest intention to enter this field until Yash Hiromi sent me some beautiful pictures (fig. 10.1) of transgenic embryos expressing β-galactosidase in the nervous system under the control of the neurogenic element of the *fushi tarazu* gene. At the time, I was spending a short sabbatical at Stanford University with David Hogness. The pictures

Hiromi sent showed a bilaterally symmetrical pattern of blue balls, presumably neuroblasts (neuronal precursor cells), in the developing nerve chord (fig. 10.1). Neither he nor I were in a position to identify these blue balls, but Stanford was an ideal place for this purpose because Corey Goodman, the expert on the insect nervous system, was working there in the Biology Department, next door to Richard Scheller, the expert on *Aplysia.*

And so one morning I walked over to Corey Goodman's laboratory and showed him the pictures. This was the beginning of a collaborative project and a long friendship. Before working on fruit flies, Goodman had worked on a "real" insect, as Carol Williams used to say, the grasshopper, where the nerve cells are big enough to allow neuroanatomical studies. The insect nerve system is built in a rather stereotypical manner, and almost every cell can be identified individually. With this knowledge of the anatomy of the grasshopper nervous system, Goodman and his graduate student Chris Doe ventured into *Drosophila.* It was hard to identify the blue balls from the photographs, but when Yash Hiromi sent them some of the original embryo preparations, they found out that *fushi tarazu* is expressed, for example, in two pairs of identified neurons (nerve cells) called anterior and posterior corner cells (aCC, pCC), and RP1 and RP2, respectively. Because homeobox genes are primarily involved in determining cell fate, we asked ourselves whether *fushi tarazu* was required for the determination of the developmental fate of these neurons and whether there would be a change of cell fate if *fushi tarazu* was not expressed in these cells. But we couldn't simply take a *fushi tarazu* mutant to examine this question, because the loss-of-function mutations primarily affect the pattern of the body segments, which has major secondary effects on the development of the nervous system. So we conceived a new genetic approach, based on Yash Hiromi's analysis of the upstream regulatory elements of the *fushi tarazu* gene.

Hiromi constructed a transgene supplied with the upstream enhancer and the zebra element but lacking the neurogenic element. The P-element containing this transgene was introduced into normal embryos and then crossed into a *fushi tarazu*–mutant background. Such embryos undergo almost normal segmentation but do not express any detectable fushi tarazu protein in the nervous system and die before reaching the second larval stage. The fate of these neurons, which in the wild type express the *fushi tarazu* gene, can then be examined by intracellular injection of the fluorescent dye Lucifer yellow, if—and only if—you are as skilled as Chris Doe, who did the injections. The yellow dye fills both the cell bodies and the axons (the fibers) and reveals the wiring in fine detail. The morphology of the four neurons (aCC, pCC, RP1, and RP2) in the normal (wild-type) embryo is shown in figure 10.2.

To our great dismay, the aCC, pCC, and RP1 cells were not affected by the absence of fushi tarazu protein. In contrast, the RP2 neuron did not behave normally. In wild-type embryos, the RP2 exon extends anterior and then laterally out along the anterior intersegmental nerve on the same side of the body, whereas the axons of many neighboring neurons such as RP1 grow

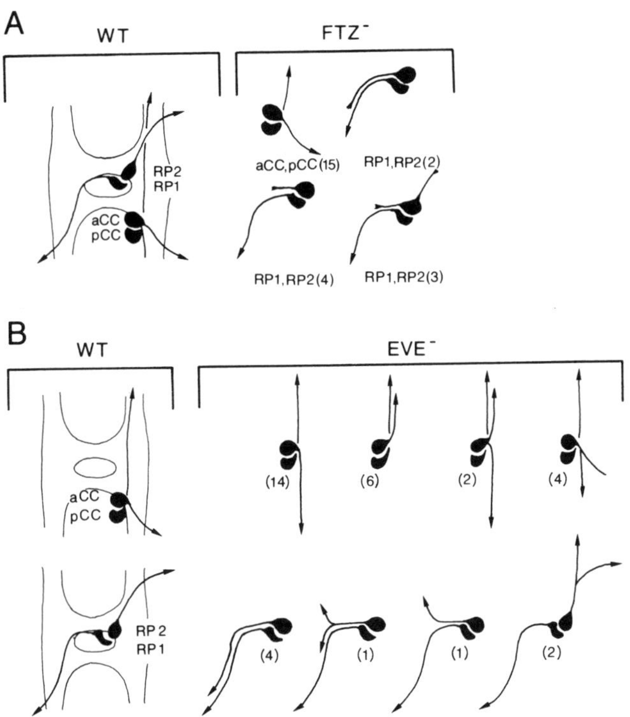

Figure 10.2
Effects of fushi tarazu *and* even-skipped *mutations on the axonal outgrowth of specific neurons. Individual neurons were filled with Lucifer yellow dye. (A) Outgrowth of aCC, pCC, RP1, and RP2 neurons in wild-type (WT) and* fushi tarazu *(FTZ⁻) mutant embryos. In FTZ⁻, aCC, pCC, and RP1 behave normally, whereas growth cones (tips) of RP2 are misrouted and behave similarly to RP1. From C. Q. Doe, Y. Hiromi, W. J. Gehring, and C. S. Goodman, Expression and function of the segmentation gene* fushi tarazu *during* Drosophila *neurogenesis,* Science *239 (1988): 170–75. (B) In* even-skipped *(EVE⁻) mutant embryos, RP2 is also affected and in addition the anterior corner cell (aCC) is misrouted. From C. Q. Doe, D. Smouse, and C. S. Goodman, Control of neuronal fate by the* Drosophila *segmentation gene* even-skipped, Nature *333 (1988): 376–78.*

across the midline, over to the other side, and then posterior and out along the posterior intersegmental nerve. In the absence of *fushi tarazu,* none of the nine injected RP2 cells extended their axon normally. Some of the axons' growth cones followed the pathway of their neighbor, RP1, and some formed two growth cones, one following RP1, the other extending anteriorly (fig. 10.2). In the absence of *fushi tarazu,* then, RP2 is transformed toward its neighbor, RP1. This was the first demonstration of the fundamental importance of homeobox genes for the correct wiring of the nervous system.

even-skipped, another homeobox gene affecting both segmentation and the development of the nervous system, is also expressed in aCC, RP1, and RP2 (but not in pCC), and later experiments by Chris Doe and Corey Goodman showed that *fushi tarazu* presumably acts through *even-skipped,* which is turned on in RP2 by *fushi tarazu.* Inactivation of *even-skipped,* acting downstream of *fushi tarazu,* also affects RP2 determination. In addition, *even-skipped* affects the determination of the anterior corner cell independently of *fushi tarazu.*

How are the instructions given by homeobox genes implemented? Since the turn of the century, when Spanish scientist Santiago Ramón y Cajal discovered neurons, the synapses of communication between them, and the growth cones that connect them up, neurobiologists have struggled to find the mechanisms of path finding of axons and growth cone guidance. In *Drosophila* you can make use of all the genetic tools to tackle this problem, and Akinao Nose, a postdoctoral fellow in Corey Goodman's laboratory, was using the enhancer detector approach just at the time when I visited Goodman again after he moved to Berkeley. With great perseverance, Nose had had screened 11,000 enhancer detector lines for expression of β-galactosidase in specific muscles. The body wall musculature of *Drosophila* embryos and larvae consists of thirty individually identified muscle fibers in each abdominal half-segment that are innervated by either one or only a few motoneurons. Nose was looking for genes expressed on the surface of single muscle fibers that might serve as an address label for the outgrowing axon to find the muscle and innervate it (supply it with nerves) properly. He did not find any genes expressed on single muscles, but two of the detector lines were expressed in a subset of muscles, as I would have predicted. To my mind, it is inconceivable that each of the thirty muscle fibers would have its own specific surface marker molecule, recognized by a single receptor expressed specifically on the surface of the growth cone of the axon which had to innervate the muscle. Similar to the reasoning about the combinatorial interactions in gene regulation, involving a combination of activators and repressors, I would argue for a combinatorial code of cell adhesion, attraction, and repulsion molecules that compose the address or zip code of a target cell.

This picture was gradually emerging in Akinao Nose's experiment. He had already cloned the gene that turned out to be *connectin*, the same gene Rob White had identified as a target gene of *Ultrabithorax*. Nose had expressed the connectin protein in bacteria and obtained antibodies against it. He reacted the antibodies with a preparation of embryos overnight, and the following morning we looked at immunostaining of the embryos. The preparations revealed a beautiful pattern of staining. The antibodies detected *connectin* exactly on the surface of those eight muscle fibers that also expressed β-galactosidase in the detector line, but *connectin* was also expressed on the axons and growth cones of those motoneurons that innervate these muscles and on several associated glial cells (fig. 10.3). During

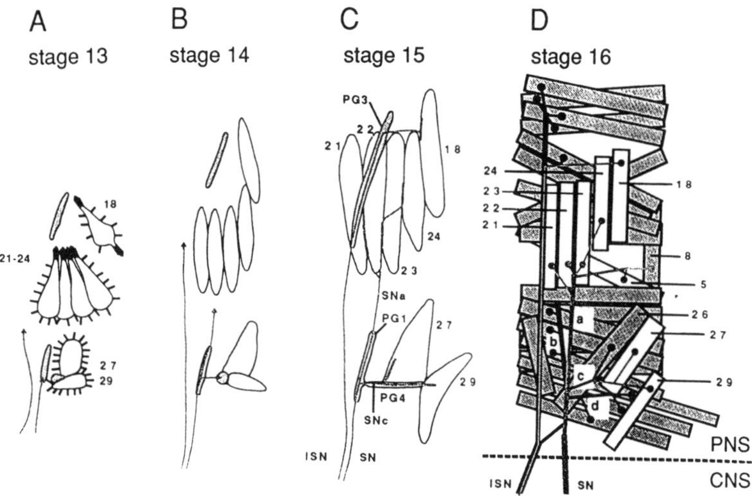

Figure 10.3
The role of connectin in nerve-muscle recognition during embryogenesis. (A) stage 13: the axons and growth cones of two outgrowing motoneurons (arrows), the pioneer muscle cells 18, 21–24, 27, and 29, and two glial cells express connectin; (B) stage 14; (C) stage 15: the motoneurons originating from the intersegmental (ISN) and segmental (5N) grow along glial cells PG3, PG1, and PG4 and contact specific muscles; (D) stage 16: the two motoneurons innervate muscles 18, 21–24, 8, 27, and 29 and form endplates (black dots). All these cells specifically express connectin. PNS peripheral nervous system, CNS *central nervous system. From A. Nose, V. B. Mahajan, and C. S. Goodman, Connectin: A homophilic cell adhesion molecule expressed on a subset of muscles and the motoneurons that innervate them in* Drosophila, Cell *70 (1992): 553–67.*

synapse formation, *connectin* localizes to the synaptic sites, where the axon contacts the muscle, but once the proper connection is made, *connectin* largely disappears. These findings immediately suggested a homophilic interaction between growth cone and muscle surface. Subsequent in vitro experiments showed that the expression of *connectin* on the surface of cells carrying different markers mediates their aggregation in a homophilic fashion. Taken together, these studies led to the suggestion that *connectin* functions as an attractive signal in axonal path finding. Later studies, however, showed that *connectin* also has a repulsive function, albeit in a different context. If *connectin* is ectopically expressed in another subset of muscles innervated by different motoneurons, their growth cones change both in their morphology and in their trajectory when they encounter their *connectin* expressing muscles. They "bypass," "detour," and "stall" instead of forming

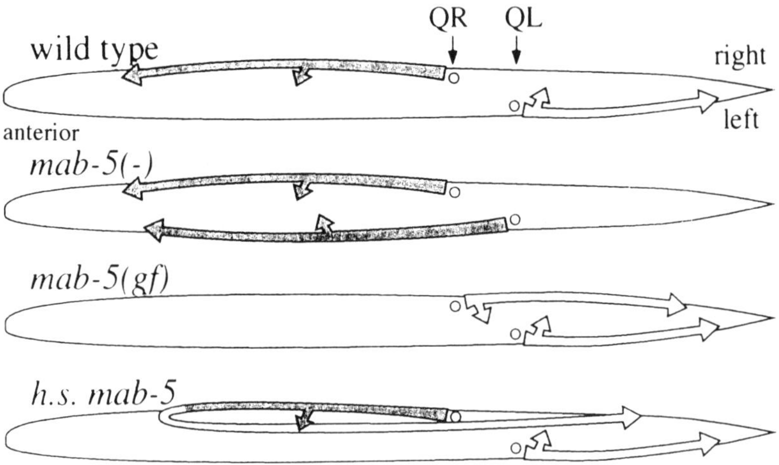

Figure 10.4
Specification of positional information by homeotic genes in the nematode worm Caenorhabditis elegans. *In the wild-type neuroblast, QR migrates anteriorly on the right side, whereas QL migrates posteriorly on the left. In* mab-5 *loss-of-function mutants (–) both neuroblasts migrate anteriorly, whereas in* mab-5 *gain-of function* (gf) *both migrate posteriorly. Using a heat-shock construct (*h.s. mab-5*) mab-5 expression can be induced after QR has migrated in the anterior direction. In this case (bottom panel), QR turns around and migrates toward the hind end, finally reaching a similar position as in the gain-of-function mutant. After S. J. Salser and C. Kenyon, Patterning* C. elegans: *Homeotic cluster genes, cell fates, and cell migrations,* Trends in Genetics *10 (1994): 159–63.*

synapses. These results clearly reveal a repulsive function for *connectin* during motoneuron growth cone guidance and synapse formation. Akinao Nose, now in his own laboratory in Japan, has since shown that *connectin* serves both a repulsive and an attractive role in synaptic targeting.

One of the most elegant experiments to demonstrate the control of the wiring of the nervous system by homeotic genes was carried out by Stephen Salser and Cynthia Kenyon in the nematode *C. elegans,* in which cell migration can directly be followed under the microscope. In the worm as in *Drosophila* most cells are born near their final position. However, some long-range migration does occur; the larval Q neuroblasts and their descendants, for example, migrate to positions spanning the antero-posterior axis. Their migration is controlled by homeotic genes that act in a cell-autonomous manner; they function within the moving cell to program the direction and extent of its migration. The homeotic *mab-5* and *lin-39* genes function as part of a left-right asymmetry program that positions sensory neurons along the antero-posterior axis (fig. 10.4). They are expressed in two neuroblasts, QL and QR, that are born opposite each other on the left and right side of the animal, but then QR migrates anteriorly and QL moves posteriorly (top panel). As QL migrates into the posterior region, where other cells express *mab-5,* it also switches on *mab-5,* which causes one of its descendants to stop moving and the other to migrate posteriorly. The function of *mab-5* in these migrations can be inferred from the mutant phenotypes: in *mab-5* loss-of-function mutants, the initial asymmetric migration remains normal, but both Q descendants migrate anteriorly (second panel), suggesting that *mab-5* is required for posterior migration. In contrast, in *mab-5* gain-of-function mutants, the Q descendants both express *mab-5* and migrate posteriorly (second panel). Now comes a pearl of an experiment: if *mab-5* is induced by heat shock (in a heat inducible gain-of-function construct), the QR descendants that have already migrated far in the anterior direction turn around and migrate posteriorly (bottom panel). This experiment demonstrates the plasticity of these cells and implies that guidance cues for posterior migration are distributed along the length of the body. It also illustrates the important role that homeotic genes play in reading positional information, which is crucial for wiring the nervous system.

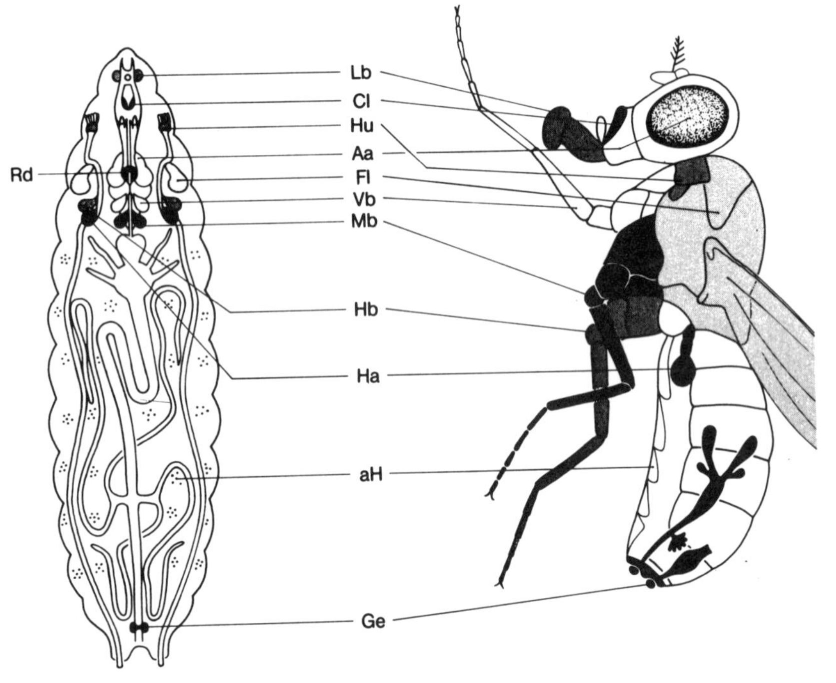

Figure 11.1

The imaginal discs and the corresponding adult structures. Abbreviations: Aa *eye-antennal disc,* aH *abdominal histoblasts,* Cl *clypeolabrum disc,* Fl *wing disc,* Ge *genital disc,* Ha *haltere disc,* Hb *hindleg disc,* Hu *humeral disc,* Lb *labial disc,* Mb *middle leg disc,* Rd *ring gland,* Vb *front leg disc. From R. Wehner and W. J. Gehring,* Zoologie, *23d ed. (Stuttgart: Thieme Verlag, 1995), fig. 3.21.*

~

11
Metamorphosis and Transdetermination

When I was a boy, my uncle Albert sent me a large cardboard box and a letter from the place where he was serving in the military. The box had small holes punctured into the cover. My mother read the letter for me. It said that the parcel contained some pupae of butterflies and that we should put them on the attic for hibernation, so that the butterflies would emerge in spring. With great curiosity I opened the box, but I saw only brown camouflaged creatures that did not move, and I thought they might have died during transportation. My mother assured me that they were not damaged, and we put them in the attic and I forgot about them. Next spring, on a fine day when the sun threw light through the small window in the roof, I opened the box. At first it was difficult to see, but then the newly emerged butterfly spread its wings and displayed its fantastic beauty. This experience transformed me into a dedicated biologist for the rest of my life.

Since then I have particularly loved flying creatures like butterflies and birds; they move so lightly through the air and display their beauty so elegantly, far surpassing earthbound creatures like us. Some creatures even mate in the air, which I consider the ultimate "savoir vivre," the zenith of evolution. My love for flying creatures is also reflected in my research topics. My first piece of research concerned bird migration. Ernst Sutter, an ornithologist from Basel, had initiated systematic radar studies on bird migration using the radar system at the Zurich airport. As a high school student I joined his team, taking photographs of the radar screen at regular intervals and filming the screen with an automatic movie camera during long nights and days in the airport control tower. What fascinated me most was the question how migratory birds navigate. Part of the accumulated data became material for my diploma thesis, which concerned diurnal migration under various weather conditions. Later, we were allowed to borrow a mobile tracking radar from the Swiss army to measure the flight route of migratory birds in the Alps. The army technicians who operated the radar told me they had been able to track eagles on this machine, but they doubted whether we would be able to detect tiny songbirds with it. But their machine was much better than they anticipated, and within a few hours I showed them that it could track individual songbirds. By looking with an oscilloscope at the "raw" signal before amplification, we could even detect the birds' wingbeat pattern. There was a telescope mounted in the axis of the radar antenna, so we could observe the flying objects both optically and with radar. When a falcon flew by, we could record its wingbeat on the oscilloscope: when it moved its wings there were waves on the oscilloscope screen and when it soared there was a straight line. Because each species or group of species of birds has a characteristic wingbeat pattern, we could assign an echo on the radar screen to certain types of birds even at night, when the migrants were not visible.

When radar was first developed during World War II, migrating birds caused several false alarms in England, especially flocks of geese, which were mistaken for German airplanes. The radar controllers called these unidentified objects "angels." We, too, had our "angels" in the Swiss Alps. They were much smaller, but the radar picked them up readily and tracked them over several hundred meters. On the oscilloscope there was a line with little wiggles, much smaller than those caused by a bird's wingbeat. When the observer stared into the telescope, he couldn't detect anything. These angels followed a straight flight path in a southwesterly direction, like the birds, but they generally flew at lower altitudes and remained invisible. But once, when I sat at the telescope, one of the angels came very near, and I saw that it was a butterfly! Butterflies and other insects also migrate across the Alps, and because the wavelength of the radar beam was below the size of the butterfly it could just detect it. To continue my studies, I would have needed access to a radar devoted entirely to research and the help of an electronics engineer. This was impossible at that time and so I turned to laboratory work for my doctoral thesis.

As mentor, I chose the leading biologist Ernst Hadorn. His biological interests were wide-ranging, and he even agreed to supervise my diploma thesis on bird migration, under the condition that I would do my doctoral dissertation in his field of research. He was then studying the development of imaginal discs of *Drosophila.* Imaginal discs, as we have seen in previous chapters, are small disclike sheets of cells in the *Drosophila* larva from which during metamorphosis the adult fly is "constructed." This brought me back to the mysteries of metamorphosis, not in butterflies, which are too delicate and difficult to study, but rather in *Drosophila,* which can be raised much more easily year-round and is quite amenable to genetic analysis. Hadorn was a fatherly figure who treated me like his intellectual son.

Hadorn had been a student of Fritz Baltzer at the University of Bern, as I mentioned in Chapter 4. In the course of his work on newt merogones, animals derived from a fertilized enucleated egg with the paternal genome alone, he became interested in the genetic control of development. After finishing his degree, he couldn't find an academic position, so he became a high school teacher and continued his research with great dedication in the cellar of his home. He realized, however, that newts were not a suitable subject for genetic analysis, and he wanted to work on fruit flies that had been introduced into genetics by Thomas Hunt Morgan. By chance, the Rockefeller Foundation asked Baltzer whether he knew of any young talented scientist who might be a candidate for postdoctoral study in the United States. Without hesitation, Baltzer recommended Hadorn, who was readily given a fellowship. He spent some time with George Beadle, one of the most outstanding students from the "fly room," who later proposed the famous "one-gene-one-enzyme theory" with Edward Tatum. Beadle gave Hadorn the first "developmental mutant" that had been isolated, *lethal giant larvae,* which provided an entry point for the study of developmental genetics. Homozygous *lethal giant larvae* cannot pupate and metamorphose; they keep crawling around in the food, accumulate hemolymph (blood) so that they become giant, and eventually die.

To find out what organ system was affected in the mutant, Hadorn designed an experiment that, as far as I know, was successful only in this case. He took *lethal giant larvae* and transplanted organ after organ from normal larvae into them, hoping to find the organ that would rescue the lethal larvae. He did not find exactly what he wanted, but when he injected a ring gland from a normal into a mutant larva, the mutant larva pupated. A fraction of the mutant larvae also pupated without intervention, but when he injected a ring gland into them, they pupated prematurely, as did injected wild-type larvae. It later turned out that the ring gland is the source of the metamorphosis hormone ecdysone. At high titers (concentrations) this hormone can induce pupation of the *lethal giant larvae.* Nevertheless, the partially rescued pupae die because of defects in tissues other than the ring gland. Many years later, Elisabeth Gateff found that *lethal giant larvae* is a tumor suppressor gene, cloned by Bernard Mechler in my laboratory. Ecdysone was purified biochemically from four hundred kilograms of silkworms by Adolf Butenandt and Peter Karlson and turned out to be a steroid hormone. The ecdysone receptor resisted biochemical isolation for many years until David Hogness and his group isolated and identified it by molecular genetic techniques. Ecdysone and its receptor are fundamentally important for the timing of gene expression during metamorphosis as well as at earlier developmental stages.

In the 1950s and 1960s, Hadorn concentrated on imaginal discs, which he studied largely by transplantation experiments. Each segment of the fruit fly larva generally has two pairs of discs, a pair of dorsal and a pair of ventral discs; the second thoracic segment, for example, has a pair of wing discs and a pair of second leg discs (fig. 11.1). During metamorphosis most of the larval tissues are broken down, and the adult fly (imago) is "rebuilt" from the building blocks, the imaginal discs. The "construction" is perfect in the sense that not the slightest trace of a discontinuity can be detected, for example, where the discs have fused, along the dorsal midline. By transplanting various imaginal discs from a donor larva to a recipient of the same stage, Hadorn showed that each disc is determined to form a precisely defined area of the epidermis with, for example, a wing or a leg. The transplanted leg would form in the body cavity of the host fly, which is filled by hemolymph and provides an ideal medium for culture. Cutting a disc into defined fragments revealed that the discs were already determined for forming defined structures within the disc and a fate map could be drawn to indicate the various primordia in each disc. I found it particularly pleasing many years later to see the fate map of the wing disc painted in blue (stained for β-galactosidase) onto the disc in an enhancer detector line reflecting the activity of a gene expressed in the various sensory organs (fig. 11.2, pl. 6).

Finally, when discs were dissociated into single cells, reaggregated, and injected into a host larva, the cells retained their state of determination. When *yellow* wing disc cells were intermixed with black (*ebony*) leg disc cells, they sorted out and differentiated autonomously. Mosaic structures consisting of *yellow* and *ebony* cells were obtained only when wing disc cells were mixed with wing disc cells or leg disc with leg disc cells, suggesting preferential adhesion of cells of the same type and/or repulsion of cells of a different type.

The behavior of disc fragments or disc cells became more complicated, however, when they were transferred into a younger host, which gave them some time for additional proliferation (growth by cell division) before the onset of metamorphosis triggered by ecdysone. In this case, the imaginal discs were capable of pattern duplication or to regenerate missing parts.

When I joined Hadorn's group it consisted of a lively bunch of people of whom I can just mention a few. There were two senior assistants, Heinrich Ursprung, my predecessor as Hadorn's research assistant, who performed hundreds of imaginal disc transplantations, and George Anders, a Dutchman. Anders was the most distracted person I have ever met. He once returned soaking wet to the institute complaining how stupid he was to have forgotten his umbrella in a thunderstorm, but he failed to notice that his umbrella was dangling from his arm. Ursprung, who shared the laboratory with Anders, considers him also the most polite person he has ever met,

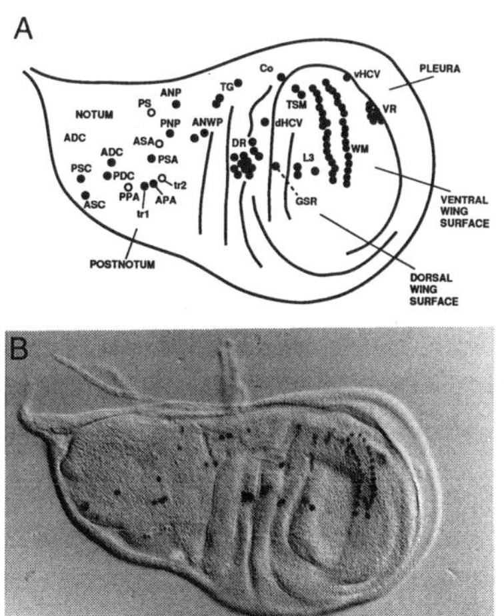

Figure 11.2
Fate map of the wing disc and localization of the various sensilla (sensory organs) in an enhancer trap line (A101). (A) Fate map. Notum, postnotum, pleura, and wing surface. The various sensilla are indicated by abbreviations. From S. Campuzano and J. Modolell, Patterning of the Drosophila *nervous system: The* achete-scute *gene complex,* Trends in Genetics *8 (1992): 202–8. (B) Precise localization of the various sensilla precursor cells in a wing disc of an enhancer trap line reflecting a gene that is expressed in the precursor cells of the respective sensilla. Courtesy of J. Modolell.*

because on several occasions he knocked on the refrigerator door before opening it. Among the students were other interesting characters—Rolf Nöthiger, Gerold Schubiger, and Heinz Tobler, while Antonio Garcia-Bellido joined the laboratory as a postdoctoral fellow. We used to argue with Garcia-Bellido for hours and hours about imaginal disc development without ever arriving at any solution of the problem, but he enjoys arguing for the sake of arguing, and these discussions, if nothing else, sharpened our intellects. One day, when I claimed that the solutions would come from molecular biology, he proclaimed, "I shall never touch a molecule," which is a difficult thing to promise.

Around that time, 1962, Hadorn and his graduate student Theo Schläpfer made an important discovery. I witnessed this discovery, and I think that it might be interesting for a reader who is not a scientist to realize how such discoveries are made. In playing around with imaginal discs Hadorn wanted to find out what happens to a disc if you skip metamorphosis and transplant it directly into an adult fly rather than a larva. The eye discs, and that was Theo Schläpfer's thesis project, would go on in the adult host and make some eye pigment, but then they would stop and not develop any further, since the ecdysone level in the adult fly is too low for further metamorphosis. Hadorn had asked Schläpfer to follow the fate of such discs if they were returned to a larval host while he did similar experiments on the genital disc. I was sharing the laboratory with Schläpfer, and besides analyzing my radar pictures, I was already doing some pilot experiments with imaginal discs. He had technical difficulties in transplanting the cultured eye discs back into larvae, so he went to ask Rolf Nöthiger for help. Nöthiger injected the eye discs that Schläpfer had recovered from the adult host back into the larvae. A week later, he dissected the host flies who had developed from the injected larvae and stared into the microscope: in addition to the beautifully red eye facets there were large pieces of wing tissue. He was completely confused. Where did this wing tissue come from? Then it dawned on him: Rolf Nöthiger must have played a trick and smuggled in some wing discs among the cultured eye discs. He asked me to act as a witness, and we went upstairs to see Nöthiger, but he claimed innocence. Theo was nervous about telling Hadorn about his results, but when he started to describe what had happened, Hadorn interrupted him and confessed that he, too, had recovered antennal and leg structures from his genital disc cultures. This was the discovery of transdetermination.

When discs proliferate by culturing in the adult host, a change in the state of determination, a transdetermination, can occur, leading, for example, from eye disc cells to wing disc cells. Almost everyone in Hadorn's crew now began to work on transdetermination. He decided to culture imaginal discs continuously in adult hosts and at each transfer to inject one sample of the tissue into a fresh adult host (as stemline) and another sample into a larva for metamorphosis (fig. 11.3). Each student was assigned a disc and the sequences in which the transdeterminations occurred were worked out.

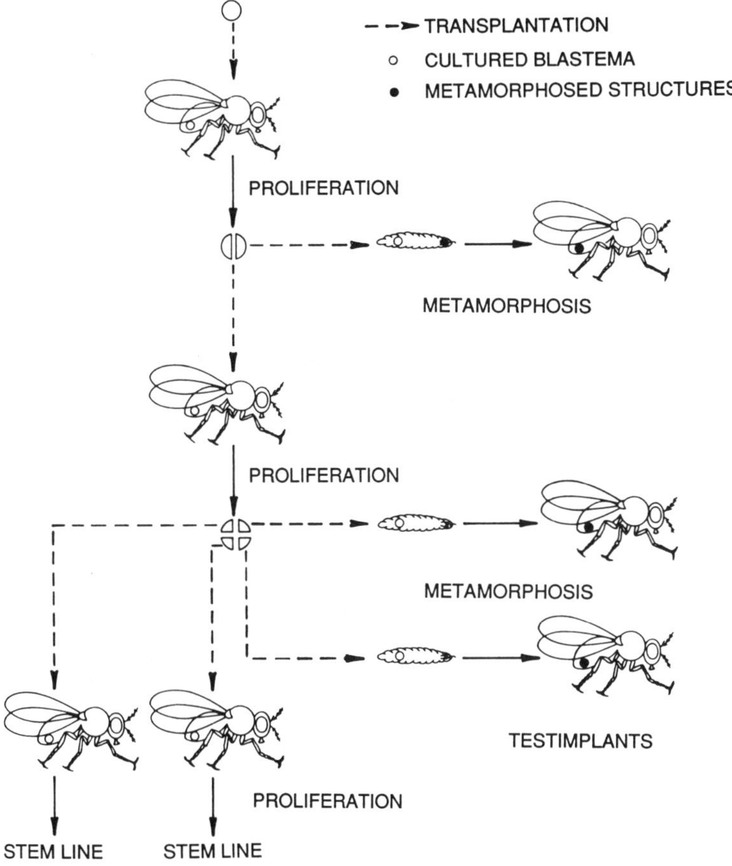

Figure 11.3
Long-term cultures of imaginal discs in vivo. The discs, or fragments thereof, are cultured in the hemolymph of adult host females by serial transplantation. After each transfer generation, part of the cultured tissue is reinjected into another female, forming the stem-line, and another fraction of the tissue is transplanted into a larval host (test implants) that subsequently undergoes metamorphosis. The metamorphosed structures can be recovered from the eclosed (hatched) adults. From H. Ursprung and R. Nöthiger, The Biology of Imaginal Discs *(Berlin: Springer Verlag, 1972), fig. 3.*

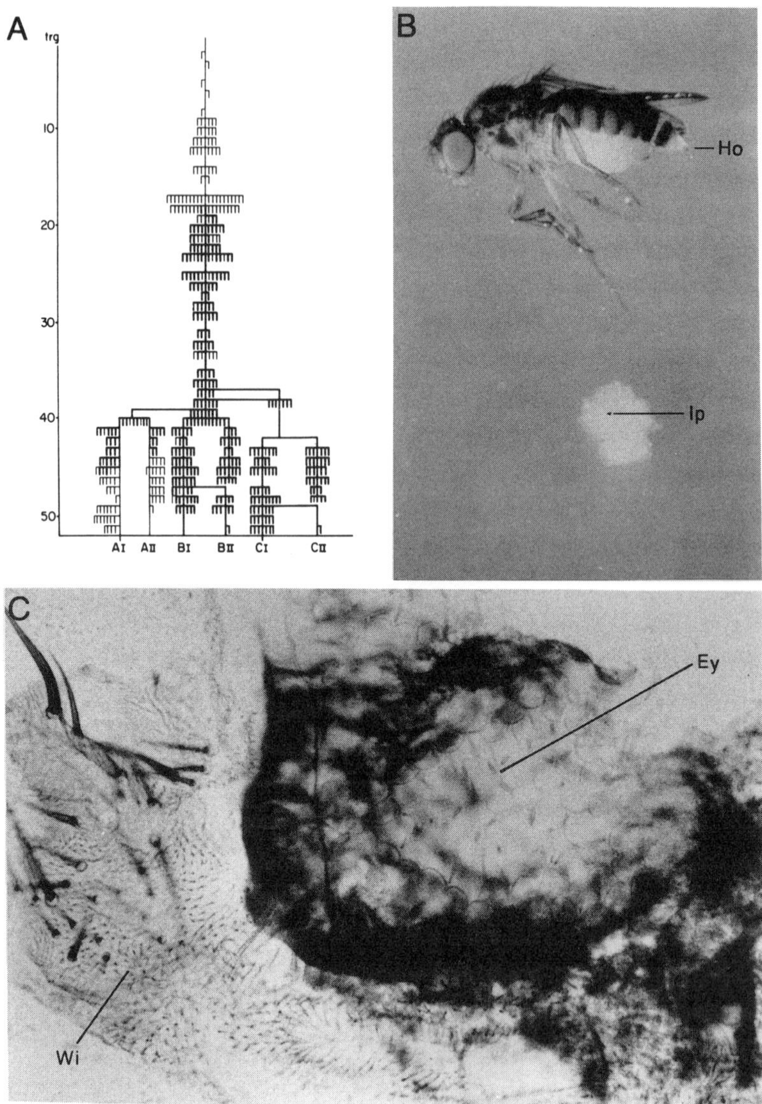

Figure 11.4
Transdetermination of imaginal disc cells determined to form wing structures to eye-forming cells. (A) Pedigree of the test implants originally derived from a cultured haltere disc. For the first seven transfer generations (trg), the test implants yielded haltere structures only. During the eighth trg a transdetermination from haltere to wing occurred spontaneously. Cells determined to form wing structures continued to proliferate for 51 trg (e.g., in subline AI). Eye-forming cells (dark line) were first observed in the 18th trg and proliferated until the 51st trg. (B) Adult host female (Ho) with implant (Ip) of imaginal disc tissue. (C) Metamorphose eye (Ey) structures with lenses, interommatidial bristles, and pigment cells, surrounded by wing cells (Wi).

Most corresponded to homeotic mutations, but there were also homeotic mutations for which no corresponding transdetermination was found, for example, mutations that converted halteres into wings and wings into halteres, though transdetermination was found to occur only in one direction and not in the other. By inducing clones of genetically marked cells in cultures of antennal discs I was able to show that transdetermination is not a clonal event but rather occurs in a group of cells. This finding makes the hypothesis of homeotic mutations occurring in the cultured disc cells as a cause for transdetermination very unlikely. A correlation was found between the extent of proliferation and the frequency of transdetermination, but at that time it was difficult to quantify the rate of proliferation in these disc cultures. It also became apparent that certain areas of the leg disc, for example, tend to undergo transdetermination at very high frequency. To this day, however, the mechanism of transdetermination remains a mystery.

One experiment I carried out as Hadorn's research assistant had a large impact on my work many years later. Hadorn had asked me and Geza Mindek, another graduate student, to study transdetermination in haltere discs. These discs are hard to culture because they grow very poorly. After a few transfer generations, the first transdetermination occurred, from haltere to wing cells. Now the wing cells overgrew the haltere cells and we could establish several sublines that produced wing cells only. These lines had lost all of the haltere cells. Then, in one of these lines a striking transdetermination occurred, from wing to eye tissue. Besides the almost transparent wing tissue, bright red eye facets were formed (fig. 11.4). This showed that under certain conditions wing cells could give rise to eye structures. We shall return to this finding in the last chapter.

As for the mechanism of transdetermination, Gerold Schubiger and I, two of the members of the Hadorn crew who have stayed in this field, are trying to solve its mysteries, now equipped with molecular genetic tools we could not even dream of at the time.

Antennapedia Homeodomain

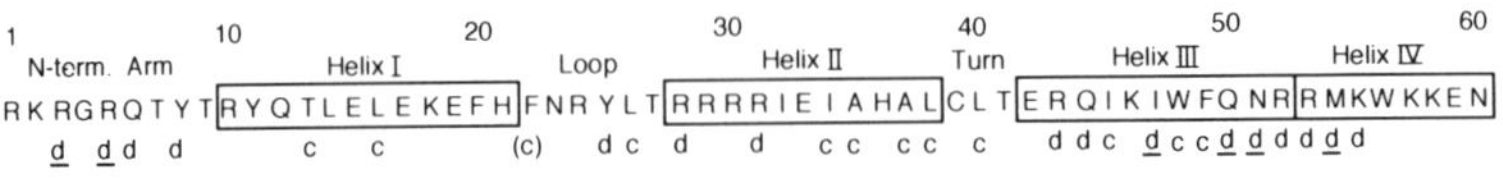

Consensus Homeodomain

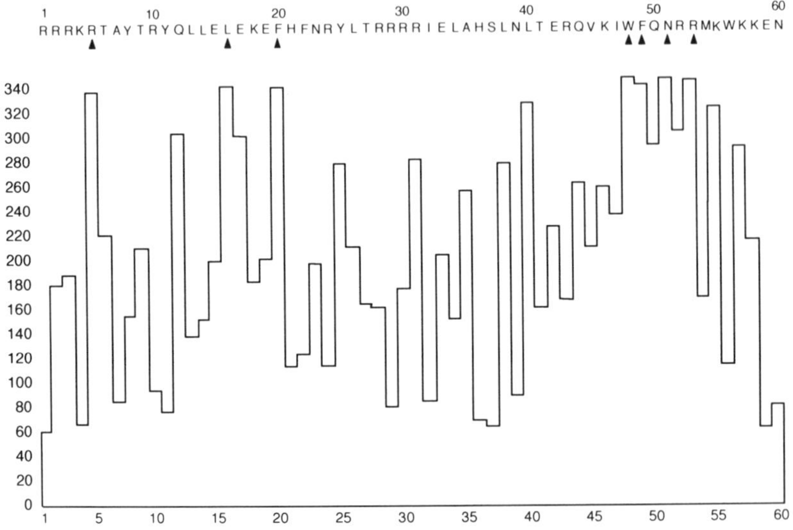

Figure 12.1
Sequence comparison between the Antennapedia *homeodomain with 346 homeodomains from* Drosophila *and other species.* Top: *Amino acid sequence (1–60), structural features; c amino acid residues involved in forming the hydrophobic core; d amino acids involved in DNA binding; d̲ amino acids making direct contacts to the DNA base pairs.* Bottom: *Consensus homeodomain and frequency diagram indicating how frequently the consensus amino acid is found at given position among 346 known homeodomain sequences from various metazoa. Black triangles indicate positions that are occupied by the same amino acid in more than 95 percent of the sequences.*

~

12
The Role of Homeotic Genes in Evolution

DNA IS a fascinating substance: the sequence of its base pairs contains the genetic information that is passed on from generation to generation. It is very much like a written text in that the sequence of its letters determines its content. On one hand, the genetic text contains a detailed developmental program as I have explained in earlier chapters, and on the other, DNA contains a historical record, information about how organisms have evolved. It thus adds considerably to the fossil record, which is filled with gaps and confined to those organisms whose remnants have been preserved over eons. Mutations that have occurred in the course of evolution are written into the genomes of living organisms. Unfortunately, mutations are superseded by secondary changes, by back mutations and deletion of genetic material, so that the historical information is incomplete and difficult to decipher. However, the evolutionary changes in DNA can be quantitated more readily than, for example, morphological characters.

After its discovery in *Drosophila,* the homeobox has been found in metazoa ranging from sponges to vertebrates and humans, in fungi and in plants, and a survey in 1995 found 346 known homeodomain sequences. The

Antennapedia homeodomain, whose structural features are indicated here, is considered the prototype (fig. 12.1). The consensus diagram indicating which amino acid occurs most frequently at a given position shows that there are seven positions (black triangles) that invariably are occupied by the same amino acid in more than 95 percent of the 346 sequences. They reveal the structural and functional constraints imposed on the molecules that are subjected to the strongest selective pressure. These include on one hand leucine (L16), phenylalanine (F20), tryptophan (W48), and phenylalanine (F49), which belong to the hydrophobic core and are essential for the proper folding of the molecule, and on the other arginine (R5), asparagine (N51), and arginine (R53) side chains, which are involved in DNA binding. Obviously, these amino acids cannot be mutated without interfering with proper folding or basic DNA binding of the homeodomain. The DNA binding specificity that differs between different types of homeodomains is determined by amino acids that are more variable, such as glutamine (Q), which is found at position 50 in the recognition helix in more than 80 percent of the known sequences but is substituted, for example, in *bicoid* and *goosecoid* by lysine or in the *Pax* genes by serine. The amino acids that form the highest pillars in this diagram define the homeobox. Very similar frequency diagrams are obtained when the sequences are plotted from a single species such as *Drosophila, Mus,* or *Homo sapiens* for which a significant fraction of the number of homeodomains have been sequenced. This indicates a very high degree of evolutionary conservation of the amino acid sequence. This conclusion is also supported by the comparison of the *Antennapedia* homologues from various species (see fig. 3.6): there is only a single amino acid difference between the homeodomain sequence of the *Drosophila Antennapedia* gene and the most closely related human gene even though vertebrates and invertebrates have evolved separately for more than 500 million years.

As more and more mammalian homeoboxes were isolated and sequenced, particularly thanks to the work of Edoardo Boncinelli and his collaborators on human genes and that of Frank Ruddle and his group on mouse genes, it became apparent that, similar to *Drosophila* homeotic genes, mammalian *Hox* genes were clustered. The homeobox sequences at the 3' end of the mammalian clusters resembled those of *labial* in *Drosophila,* whereas those at the 5' end showed a similarity to the homeobox of *Abdominal-B,* suggesting an arrangement of the *Hox* genes similar to that found for the homeotic (*Hom*) gene complexes of *Drosophila.* In addition, the direction of transcription of all the genes in the *Hox* cluster is the same

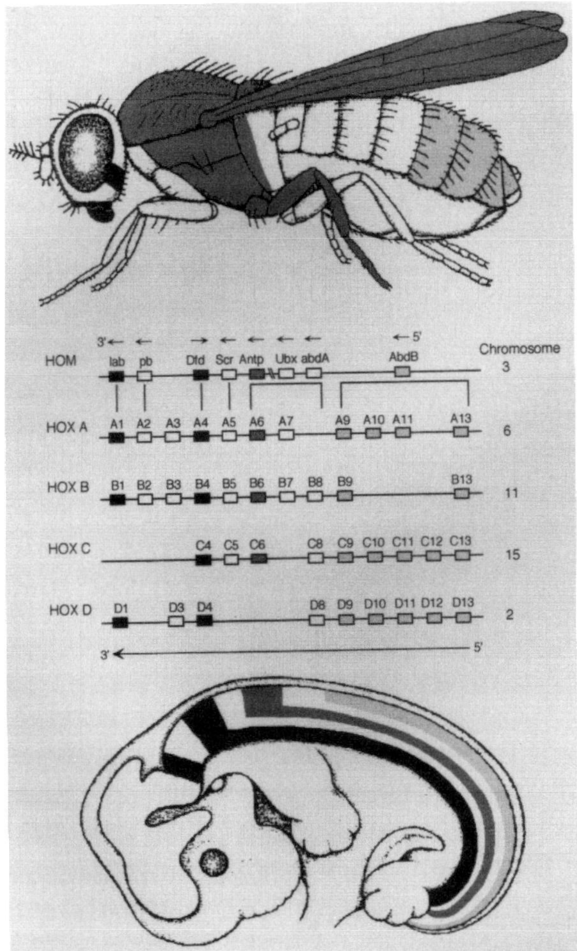

Figure 12.2

Colinearity of chromosomal arrangement of homeotic genes and their expression along the antero-posterior body axis. Top: Drosophila *schematic illustration of the expression pattern and the homeotic Antennapedia and Bithorax complexes on chromosome 3 of* D. melanogaster. *The direction of transcription (5'–3') is indicated by arrows. The most anteriorly expressed genes are localized at the 3' end, and the most posteriorly expressed genes are localized at the 5' end of the cluster.* Bottom: *The mouse has four paralogous clusters on different chromosome HoxA-D. Paralogue groups 6, 7, and 8 are very similar and cannot be assigned to a corresponding* Drosophila *gene with certainty. Paralogue groups 9–13 all correspond to* Abdominal-B *in* Drosophila. *All genes in a cluster are transcribed in the same orientation. The expression patterns in the mouse embryo are illustrated schematically underneath. From E. M. De Robertis, G. Oliver, and C. V. E. Wright, Homeobox genes and the vertebrate body plan,* Scientific American *263 (1990): 26–33.*

as it is in *Drosophila*, with *Deformed* being the sole exception. In situ hybridization experiments carried out by Steven Gaunt in collaboration with Denis Duboule and subsequently by Robb Krumlauf and his co-workers showed not only that the chromosomal organization was conserved between *Hom* and *Hox* but that the same rule that more anteriorly expressed genes were located at the 3' end of the cluster and the more posteriorly expressed genes toward the 5' end also applies to mammals (fig. 12.2). This remarkable colinearity of chromosomal organization and expression is so far unique to this group of genes. It is a strong argument in favor of the hypothesis that the *Hox* and *Hom* genes are true homologues—that is, orthologues related by descent.

Vertebrates and invertebrates differ in one significant respect here: vertebrates possess four *Hox* clusters (as indicated in fig. 12.2 for the mouse) on four separate chromosomes, as compared to a single cluster in insects. This is the result of partial genome duplications that occurred in the course of vertebrate evolution, as reflected from the genetic mapping data accumulated by Frank Ruddle and others. Peter Holland and Jordi Garcia-Fernandez have shown that the lancelet (*Amphioxus*), one of our earliest

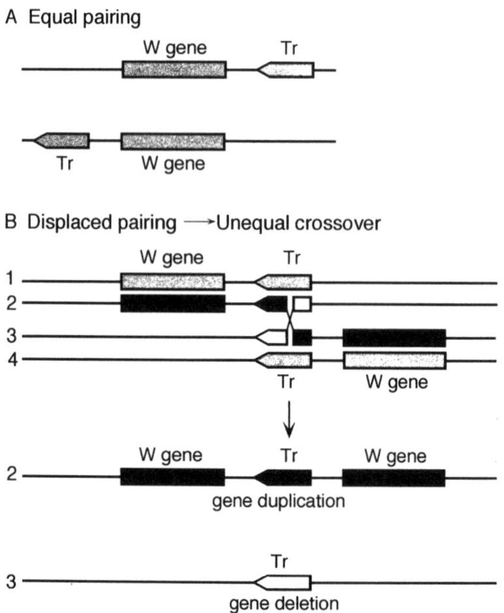

Figure 12.3
Gene duplication by transposon-induced unequal crossover: (A) equal pairing at the white locus (w) with two transposons (Tr) inserted in the same orientation left and right of the white *gene respectively in the two homologous chromosomes; (B) displaced pairing and homologous recombination between the two transposons result in a duplication and a deletion of the* white *gene.*

known "ancestors," still has just one *Hox* cluster. A similar cluster, albeit with fewer genes, has been isolated from the nemertean worm *Lineus* by Marie Kmita-Cunisse in my laboratory. Remnants of a cluster have been found in *C. elegans,* and a cluster may even be present in cnidarians (jellyfish, sea anemones, and other coelenterates), but in sponges only different types of homeobox genes have been found so far, and no *Antennapedia*-type genes. Denis Duboule has proposed that the colinearity of chromosomal organization also applies to the temporal order in which the genes are expressed, and experiments in which he switched the position of the *Hoxd-9* gene within the cluster are consistent with this hypothesis.

How could the homeotic gene cluster have evolved, and what are the selective forces that keep the genes clustered? The answers to these questions have a long history that Ed Lewis reviewed in 1992. We still have only partial answers, but molecular data have added considerably to the evidence and lead to interesting speculations that I should like to present here.

In 1925 Alfred Sturtevant discovered the first case of a gene duplication, the famous *Bar*-eyed mutant. The progeny of homozygous *Bar* flies revert at an unusually high frequency to (normal-eyed) wild type and mutate to (tiny-eyed) *Ultrabar.* These two exceptional types of progeny were associated with an unusual type of recombination, which he called unequal crossover. He interpreted this finding as displaced pairing of a duplicated gene and crossover leading from duplication (two copies of the gene) to triplication (three copies) and back to the wild type (one copy). In 1936, after the giant polytene chromosomes were discovered, Calvin Bridges and Hermann Muller independently confirmed Sturtevant's basic idea by showing that the unequal crossover at the *Bar* locus results from a small tandem duplication of seven chromosomal bands. It was harder to explain how the gene duplication first arose. With the help of Mel Green, who had isolated putative duplications of the *white* gene many years earlier, Michael Goldberg in my laboratory showed that transposons can cause gene duplications. Transposons can insert essentially at random positions in the genome and provide the homologous DNA sequences for recombination. If two transposons insert left and right of the *white* gene or any other target gene and are arranged in the same orientation (fig. 12.3), the homologous chromosomes can pair unequally, and recombination leads to a duplication and the reciprocal deletion of the *white* gene. The molecular data that we obtained are entirely consistent with this interpretation. Once the gene is duplicated, it can easily be triplicated as found in *Bar* and a linear array of multiple genes can be generated by successive rounds of unequal crossover. These findings form the basis for the model presented below.

Ed Lewis was first led to a model of gene duplication by unequal crossover at the *bithorax* locus by observations of Calvin Bridges, who interpreted certain types of salivary gland chromosome banding patterns as the result of gene duplications, but genetic evidence for such duplications was lacking. The *bithorax* locus maps to two doublets in section 89E of the third chromosome and was a candidate for such gene duplications. In his genetic studies, Lewis arrived at a model of the Bithorax complex with one gene specifying each segment that might have arisen by tandem duplication. However, the subsequent genetic analysis by Gines Morata and the molecular cloning of the complex showed that there were only three protein coding genes with one homeobox each. Nevertheless, the clustered arrangement of the homeotic genes in tandem and in the same orientation strongly suggests unequal crossover as a mechanism of cluster formation. A possible model for the generation of the homeotic gene clusters is presented in figure 12.4. It is based on the findings that *Antennapedia* and its cognate *Hox* genes (*Hox6* and *Hox7*), located in the center of the cluster, deviate the least from the consensus sequence, and that the sequences progressively diverge toward more

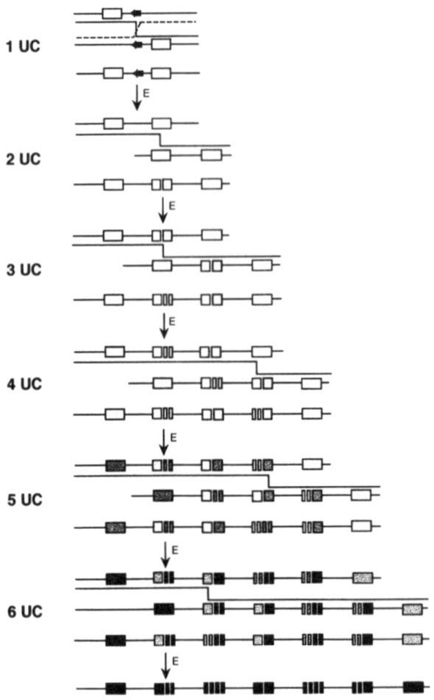

Figure 12.4
Model for the generation of a gene cluster by unequal crossover. The first unequal crossover (1 UC) occurs between two transposons. The second (2 UC) and subsequent unequal crossovers occur between the duplicated genes. Because displaced pairing is a prerequisite for each evolutionary duplication event (E), the outside flanking genes are not affected, whereas the interior genes are progressively homogenized.

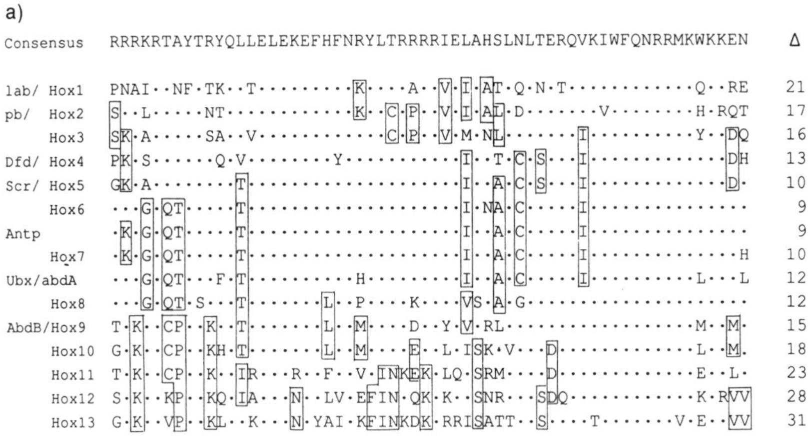

a)

```
Consensus    RRRKRTAYTRYQLLELEKEFHFNRYLTRRRRIELAHSLNLTERQVKIWFQNRRMKWKKEN   Δ

lab/ Hox1    PNAI··NF·TK··T·········K····A··V·I·AT·Q·N·T············Q··RE  21
pb/  Hox2    S··L·····NT············K··C·P··V·I·AL·D·······V········H·RQT  17
     Hox3    SK·A·····SA··V············C·P··V·M·NL·······I··········Y··DQ  16
Dfd/ Hox4    PK·S······Q·V········Y···········I··T·C·S···I·············DH  13
Scr/ Hox5    GK·A········T····················I··A·C·S···I·············D·  10
     Hox6    ···G·QT·····T····················I·NA·C·····I···············   9
Antp         ·K·G·QT·····T····················I··A·C·····I···············   9
     Hox7    ·K·G·QT·····T····················I··A·C·····I··············H  10
Ubx/abdA     ···G·QT···F·T··········H·········I··A·C·····I··········L···L  12
     Hox8    ···G·QT·S···T·······L··P····K····VS·A·G·····················  12
AbdB/Hox9    T·K··CP··K··T·······L··M····D··Y·V·RL··················M··M·  15
     Hox10   G·K··CP··KH·T·······L··M····E··L·ISK·V···D·············L··M·  18
     Hox11   T·K··CP··K··IR···R··F··V·INKEK·LQ·SRM····D·············E··L·  23
     Hox12   S·K··KP··KQ·IA···N··LV·EFIN·QK·K··SNR···SDQ············K·RVV  28
     Hox13   G·K··VP··KL··K···N·YAI·KFINKDK·RRISATT··S····T·······V·E··VV  31
```

Figure 12.5

Consensus homeodomain sequences of the homeodomain families and paralogue groups. Amino acids differing from the consensus sequence on the top that are shared by adjacent families are boxed. Δ indicates the number of amino acids that differ from the consensus sequence. From W. J. Gehring, M. Affolter, and T. Bürglin, Homeodomain proteins, Annual Review of Biochemistry *63 (1994): 487–526.*

anterior or more posterior genes, respectively (fig. 12.5). Finally, the terminal genes *labial/Hox1* and *Abdominal-B/Hox13* contain the most divergent homeodomains. This suggests that *labial/Hox1* and *Abdominal-B/Hox13* are the oldest genes, with the longest time to diverge from the original sequence. The first duplication thus presumably led to the two terminal genes of the cluster, and a series of consecutive unequal crossovers then generated the interior genes. Due to these multiple recombination events that do not affect the terminal genes, the internal genes are mosaics, combining segments of their ancestral genes, and their sequences become homogenized. Because the internal genes were generated later in evolution, they have diverged less from the primordial gene than the terminal genes. Supporting this hypothesis is the finding that the terminal genes *labial, proboscipedia,* and *Abdominal-B* share an intron at the same position in the homeodomain (between amino acid 44 and 45). Missing, however, are data on earlier stages of evolution of the cluster, when fewer genes were present. The nemertean worm *Lineus* may carry such a cluster, but it is difficult to prove that a certain gene has not secondarily been lost. This model provides a partial explanation for the colinearity rule only; much more work is required before we can reconstruct the sequence of events.

At least two genetic mechanisms seem to be involved in maintaining the chromosomal order of the genes. For several pairs of neighboring genes in both mouse and *Drosophila* there is evidence that they share the same cis-regulatory region or enhancer. This leads to a functional linkage of the two genes, because translocation or inversion events separate one or the other gene from its cis-regulatory region or enhancer and are therefore detrimental. Nevertheless, the homeotic gene cluster may occasionally be split, as in *Drosophila.* Interestingly, the split has occurred between *Antennapedia* and *Ultrabithorax* in *Drosophila melanogaster,* whereas in *D. virilis* the cluster is split between *Ultrabithorax* and *abdominal-A,* as François Karch and his collaborators discovered. Furthermore, a gene in the cluster may become inverted, as for example *Deformed.* Dado Boncinelli and co-workers have

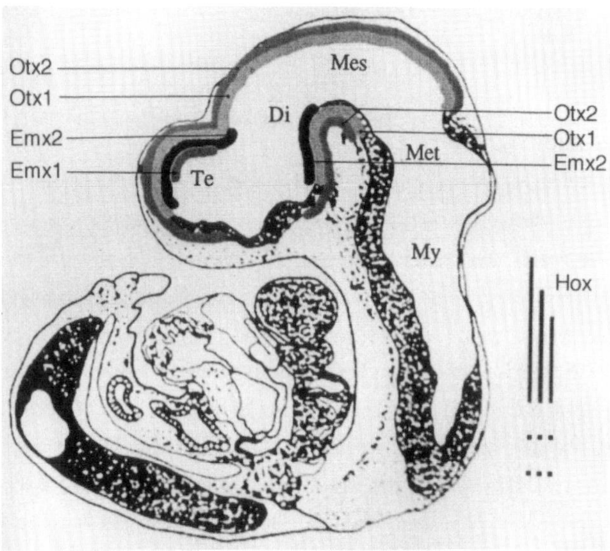

Figure 12.6
*Expression pattern of the most anteriorly expressed homeobox genes (*Otx1 *and* Otx2, Emx1 *and* Emx2*) in the forebrain of the mouse.* Abbreviations: Di *diencephalon,* Hox *levels of expression of the three most anteriorly expressed* Hox *genes,* Mes *mesencephalon,* Met *metencephalon,* My *myelencephalon,* Te *telencephalon. After A. Simeone, D. Acampora, M. Gulisano, A. Stornaiuolo, and E. Boncinelli, Nested expression domains of four homeobox genes in the developing rostral brain,* Nature *358 (1992): 687–90.*

found yet another stabilizing factor in the human *Hox* cluster, where a single 5' exon may be spliced onto the transcripts of several genes. Such a mechanism would also keep the gene order constant.

The expression patterns of *Hox* genes in the mouse (fig. 12.2) show that the expression of the anteriormost *Hox* gene does not extend into the region of the forebrain, the part of the body plan neurobiologists are most interested in. This provokes the question of which genes specify the anterior head region. Numerous homeobox genes lie outside the clusters. Some may have a different origin and may never have been part of the cluster, whereas others may have originated in the cluster and were translocated to a different chromosomal locus. The latter may be the case for *caudal* and *empty spiracles,* which in addition to their homeoboxes share a hexapeptide motif with the consensus amino acid sequence IYPWMK, which is found in all the clustered homeotic genes but *Abdominal-B.* Furthermore, they have an intron at the same position (amino acid 44/45) in the homeobox as *labial, proboscipedia,* and *Abdominal-B.* Because intron formation is a rare evolutionary event, genes that share an intron splice site are very likely related. *empty spiracles* and *caudal* may therefore have arisen from the cluster, and because they are expressed in the anterior head and the posterior tail region, they may have been located near the ends of the cluster. Antonio Simeone, knowing that *empty spiracles* and another homeobox gene called *orthodenticle* are expressed in the anterior part of the head in *Drosophila,* used these genes as probes to isolate their mammalian homologues. As he predicted, mammalian homologues are expressed in the forebrain of the mouse embryo in an expression pattern that corresponds to that in *Drosophila* (fig. 12.6). In collaboration with Katsuo Furukubo-Tokunaga and Heinrich Reichert, we have analyzed the expression pattern of *empty spiracles* and *orthodenticle* in the *Drosophila* brain and determined that the organization of the *Drosophila* and mouse brain into neuromeres specified by those two genes is surprisingly similar.

So far, we have seen that the sequences of homeoboxes, the organization of homeotic genes into clusters, and their expression patterns are very similar in *Drosophila* and mammals, but do the homeobox genes serve the same function in mammalian development as they do in *Drosophila?* To answer this question, both gain- and loss-of-function mutations were generated in the *Hox* genes of the mouse. In contrast to the *Pax* genes, which are discussed in the final chapter, only a few spontaneous or induced mutations of the *Hox* genes are known. This may be due to the partial functional redundancy

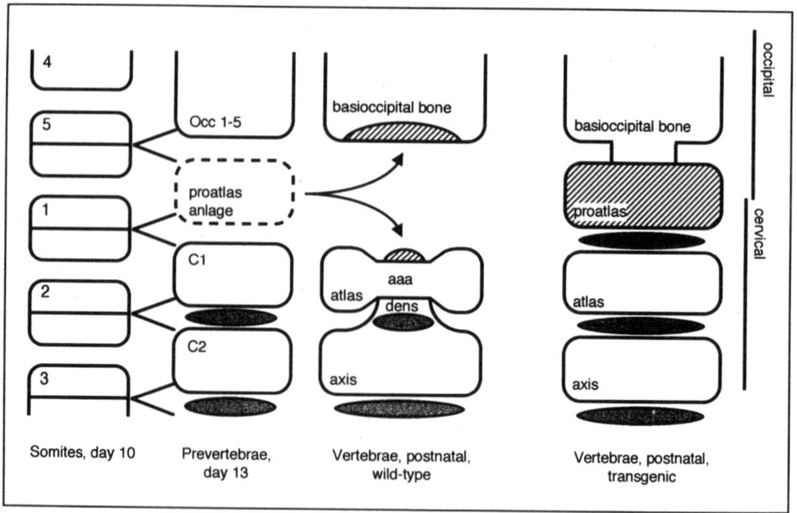

Figure 12.7
Induction of an additional cervical vertebra, a proatlas, by ubiquitous expression of the Hoxa-7 *gene in the mouse.* Left panels: *Development of the basioccipital bone, atlas, and axis in the wildtype mouse.* Occ *occipital bones,* C *cervical vertebrae.* Right panel: *Formation of a proatlas in the* Hoxa-7 *gain-of-function mutant. From M. Kessel, R. Balling, and P. Gruss, Variations of cervical vertebrae after expression of a* Hox-1.1 *transgene in mice,* Cell *61 (1990): 301–8.*

between the paralogous genes in the four *Hox* clusters, which means that the mutations had to be engineered. I shall discuss two examples to illustrate the basic effects of *Hox* mutations. The first gain-of-function mutation was constructed by Michael Kessel, Rudi Balling, and Peter Gruss in Göttingen. They fused the coding region of the *Hoxa-7* gene, not onto a heat shock promoter, as my laboratory had done previously for *Antennapedia,* but onto an actin promoter that enhances expression more or less uniformly all over the mouse embryo. Transgenic mice carrying this gain-of-function construct are born with abnormalities of the face and the skull and die after birth (fig. 12.7). They form an additional cervical vertebra, a proatlas, which is derived in part from the basioccipital bone of the skull by a homeotic transformation. The atlas and axis also assume characteristics of more posterior vertebrae in these mice. These changes indicate that *Hoxa-7* gain-of-function mutations cause posterior homeotic transformations much as *Antennapedia* does in *Drosophila.* The formation of an additional cervical vertebra represents a reversion to an earlier stage of evolution that is represented by rep-

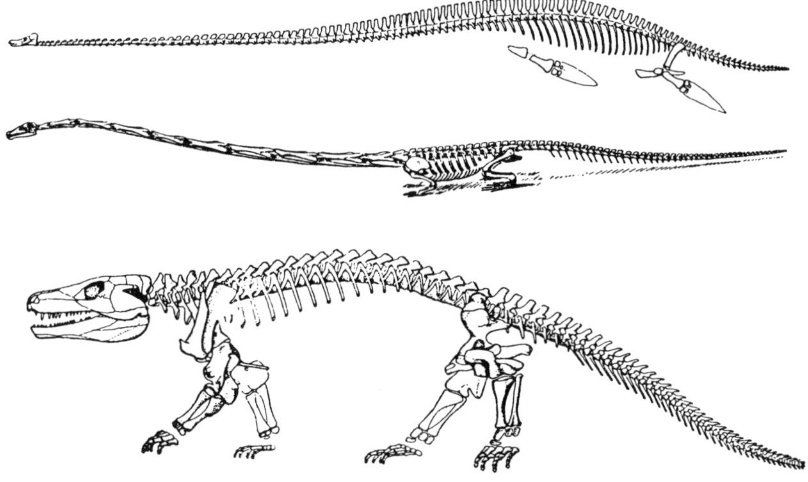

Figure 12.8
Variation of the number of cervical vertebrae in fossil reptiles. Top: Elasmosaurus *(Cretaceous), with many cervical vertebrae;* middle: Tanystropheus *(Triassic,) with fewer elongated cervical vertebrae;* bottom: Seymouria *(Permian), a primitive reptile from the transition zone between amphibians and reptiles with six cervical vertebrae. From E. Kuhn-Schnyder,* Geschichte der Wirbeltiere *(Basel: Schwabe, 1953), 89, 90.*

tiles. Proatlases have been detected in the fossil skeletons of dinosaurs and rudiments are present in living reptiles. All mammals, with rare exceptions (tree sloths and manatees), have seven cervical vertebrae, whereas this number is much more variable in reptiles (fig. 12.8), which seem to experiment with the construction of the neck: a long neck can be formed increasing either the number of cervical vertebrae or their length. The reason why all mammals (including the giraffe) have exactly seven cervical vertebrae is a result of history: mammals evolved from mammal-like reptiles that happened to have seven cervical vertebrae. In this way we can turn the wheel of evolution backward by introducing such homeotic mutations.

The reciprocal homeotic transformation was achieved by Mario Capecchi and his collaborators, who disrupted the *Hoxa-3* gene of the mouse by homologous recombination (see fig. 4.20). Mice homozygous for this loss-of-function mutation show a partial transformation of the atlas to an extension of the basioccipital bone (fig. 12.9). This indicates that gain- and loss-of-function mutants have the opposite effect and cause homeotic transformations in the opposite direction, exactly as in *Drosophila. Hox* genes therefore must be considered true homeotic genes. These experiments also confirm a hypothesis proposed by Johann Wolfgang von Goethe, who was interested in science and evolution. Goethe proposed "Die Wirbeltheorie des Schädels," the idea that in the course of evolution some bones of the skull were derived from vertebrae. This theory is strongly supported by the above experiments revealing the transformation of a vertebra into a bone of the skull.

The most direct way to test whether the *Hox* genes of the mouse can function as homeotic genes is to introduce the mouse gene into *Drosophila.* This experiment was carried out by my former associate Bill McGinnis and his coworkers by inserting the mouse *Hoxb-6* gene into a heat-shock vector and expressing it in *Drosophila. Hoxb-6* is a putative *Antennapedia* homologue. When induced by heat shock, *Hoxb-6* induces antenna-to-leg transformations in adult flies and head segment into second thoracic segment transformations in embryos (fig. 12.10). These homeotic transformations are essentially indistinguishable from those induced by the *Antennapedia* gene in *Drosophila.* As this dramatic effect clearly demonstrates, the *Hox* genes are functional homologues of the homeotic *Drosophila* genes. Leslie Pick extended these studies to the *Hoxa-5* gene, which is a homologue of *Sex combs reduced* in *Drosophila,* demonstrating that heat-shock induction of this

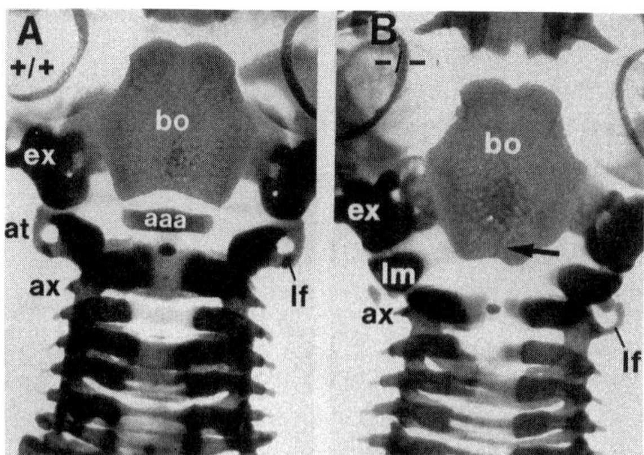

Figure 12.9
Disruption of the Hoxa-3 *gene leads to a transformation of the atlas into parts of the basioccipital bone. After B. G. Condie and M. R. Capecchi, Mice homozygous for a targeted disruption of* Hoxd-3 (Hox-4.1) *exhibit anterior transformations of the first and second cervical vertebrae, the atlas, and the axis,* Development *119 (1993): 579–95.*

gene produces the same phenotype as that of *Sex combs reduced* and confirming the gene specificity of the transformation (fig. 12.10). Pick and her collaborators also showed that *Hoxa-5* directly regulates *forkhead,* a target gene of *Sex combs reduced,* and thus participates in the regulatory hierarchy of insect morphogenesis. Insects can thus "understand" the genetic message of mice.

If the homeodomain is so highly conserved in evolution, one would

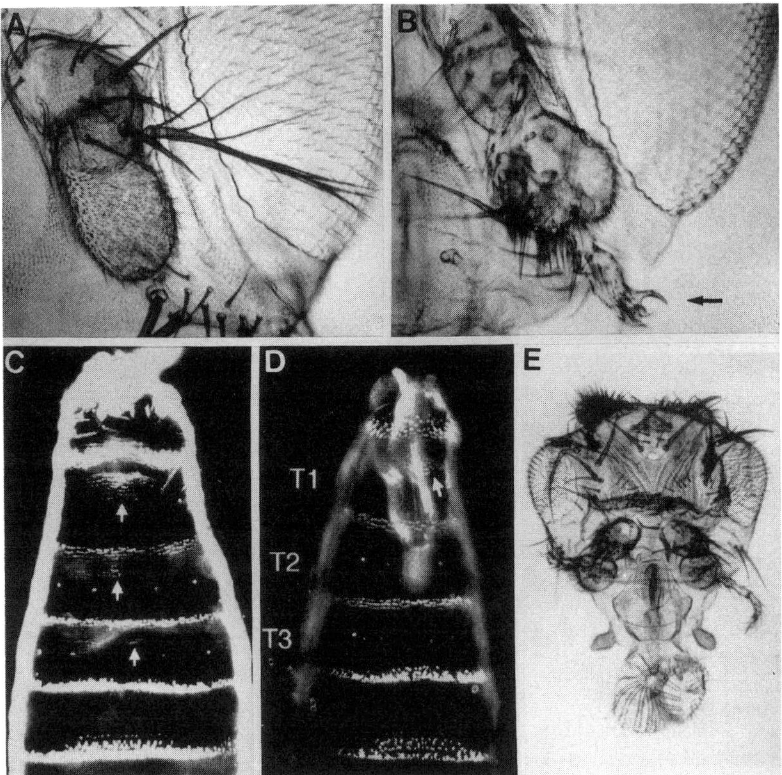

Figure 12.10
Homeotic transformations induced by mouse Hox *genes in* Drosophila. *(A) Normal antenna. (B) Antenna-to-leg transformation induced by ubiquitous expression of* Hoxb-6, *a putative* Antennapedia *homologue. (C) Homeotic transformation of T2 and T3 toward T1 (arrows are pointing to the beardlike structures that are typical for T1) induced by ubiquitous expression of* Hoxa-5. *(D) Normal embryo. T1–T3 thoracic segments. (E) Homeotic transformation of arista (most distal part of the antenna) into tarsal structures and the back of the head into dorsal thoracic structures. A–B, From J. Malicki, K. Schughart, and W. McGinnis, Mouse* Hox-2.2 *specifies thoracic segmental identity in* Drosophila *embryos and larvae,* Cell *63 (1990): 961–67. C–E, After J. J. Zhao, R. A. Lazzarini, and L. Pick, The mouse* Hox-1.3 *gene is functionally equivalent to the* Drosophila Sex combs reduced, Genes and Development *7 (1993): 343–54.*

expect that the sequences of its DNA binding sites are also highly conserved along with some of the regulatory circuits, because the homeodomain and its target genes co-evolve. In collaboration with Debra Wolgemuth, my laboratory tested this hypothesis for the *Hoxa-4* gene. The intron of this gene contains a highly conserved regulatory element that was detected by comparing the intron sequences of the homologues and paralogues of other vertebrate species. When my graduate student Theo Haerry introduced this element into *Drosophila,* he could show clearly that it is responsive to *Drosophila* homeotic genes. Flies thus seem to "understand" mouse regulatory elements.

It appears that the homeobox has revealed a universal genetic mechanism for the control of morphogenesis. The specification of body plan along the

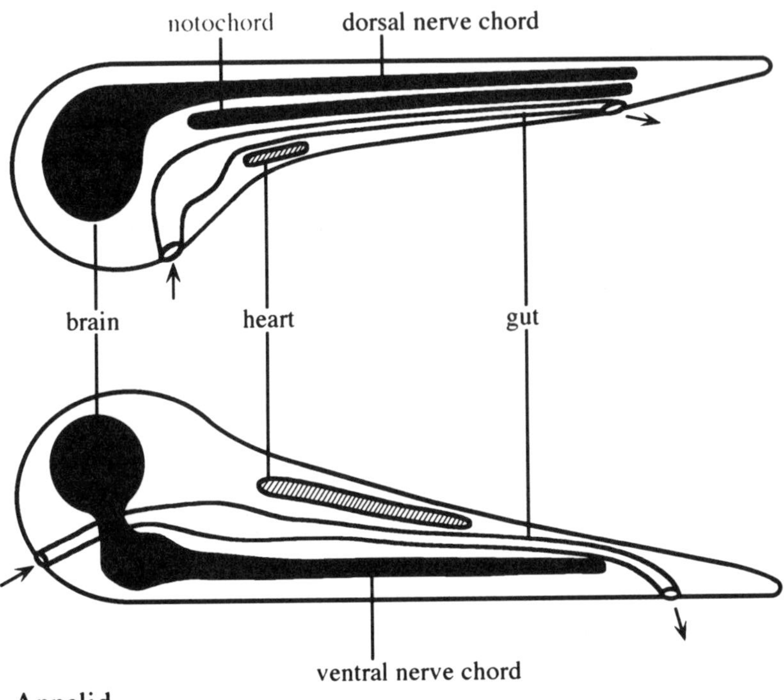

Figure 12.11
Hypothetical dorso-ventral inversion of the body plan in vertebrates as compared to invertebrates. The general body plans of a vertebrate and an annelid worm are represented schematically.

antero-posterior axis is controlled by the same homeotic genes in insects and mammals and, by extrapolation, in all the other metazoa that possess homeobox genes. Whether the same applies to the dorso-ventral axis is not immediately obvious. The positional information along the dorso-ventral axis is specified mainly by genes that do not have homeoboxes. A major difference in the basic body plan of vertebrates and invertebrates is found in the dorso-ventral axis: vertebrates have a dorsal nervous system, whereas most invertebrates have a ventral nervous system. This apparent distinction can be resolved if one assumes that vertebrates at some stage of their evolution turned on their backs, as Etienne Geoffroy Saint-Hilaire proposed in his famous dispute with Georges Cuvier. This is not an impossible assumption; the flat fish of the sole family, for example, turn on their side during development. Geoffroy Saint-Hilaire's hypothesis is supported by the expression pattern of genes involved in specifying the dorso-ventral axis. A striking example is the *decapentaplegic* gene, which encodes a signaling molecule, and its mammalian homologue, the bone morphogenetic protein-4 (BMP-4) gene. Whereas *decapentaplegic* is expressed dorsally in the early *Drosophila* embryo, BMP-4 is expressed ventrally in the mouse body plan. The number of such cases is gradually increasing, and two even more convincing examples are presented in the final chapter. And so Geoffroy Saint-Hilaire's idea that the body plan of a vertebrate can be explained simply by turning that of an annelid worm upside-down (fig. 12.11) may be correct after all.

Homeotic genes have played a fundamental role in evolution by specifying the body plan. Because they concern master control genes, homeotic mutations have dramatic effects on morphogenesis. Antennae may become legs, or a new pair of legs may be generated on the first abdominal segment. Most of these dramatic mutational effects are lethal or deleterious, but occasionally a change in the body plan is favorable and leads to new lines of evolution. Any mutation in the master control genes must be followed by many small mutational changes in its target genes further down in the regulatory hierarchy to optimize the new developmental pathway through a mechanism that might be called streamlining. To date, efforts have focused on reconstructing these evolutionary events, but in the future it might be possible to approach these evolutionary problems through experiments.

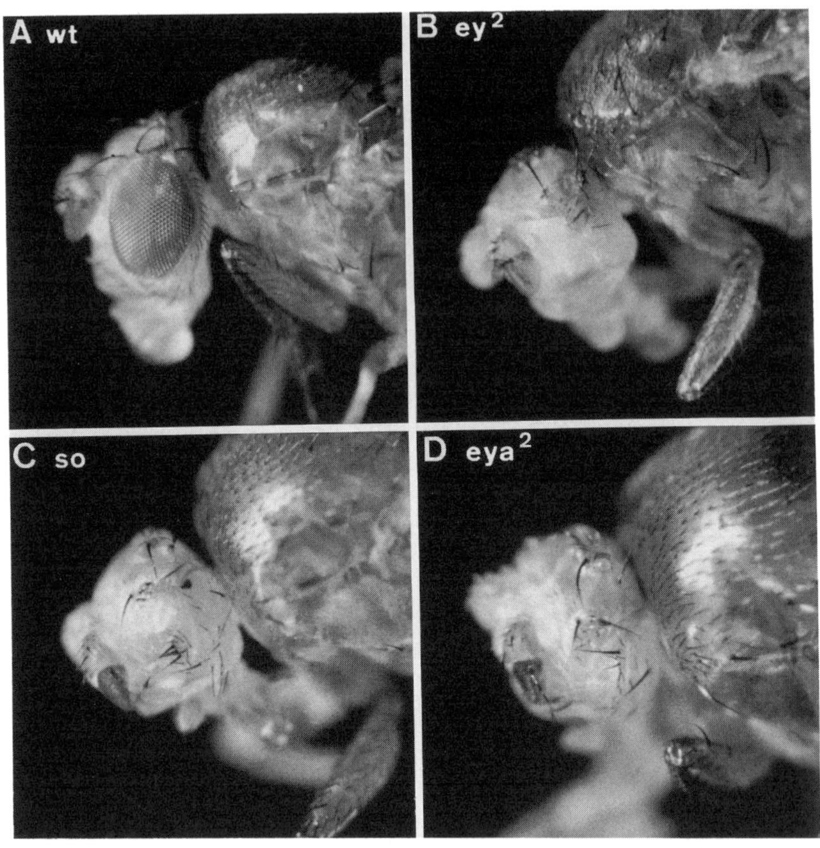

Figure 13.1
Drosophila *mutants lacking eyes: (A) wild type with normal eyes; (B)* eyeless *(*ey^{2}*);*
(C) sine oculis *(*so*); (D)* eyes absent *(*eya^{2}*). Courtesy of U. Kloter.*

~

13
A Deep Look into the Eyes

THE COMPUTER'S printer began to spit out lots of sequences, and all of a sudden I realized that we had a great discovery on our hands. Earlier that day, Rebecca Quiring had finished sequencing her gene and entered the first five hundred base pairs into the computer to search for similar sequences stored in the European Molecular Biology Laboratory's database in Heidelberg. She had thought that the answer would come back right away, but I told her that even though she had used a program called FAST A, the computer now had to search through some ten million base pairs of sequences to find similarities and this would take some time. So we went to lunch, and after returning the miracle happened: the computer had found a large number of similar sequences and ordered them with respect to the extent of similarity. The highest score was found for the mouse *Pax-6* and the human *Aniridia* genes, followed by the *Drosophila* paired gene, and it dawned on me then that Rebecca Quiring had probably cloned the *eyeless* gene of *Drosophila*. Urs Kloter had mapped her gene by in situ hybridization to section 102D on the fourth chromosome, and one of the few genes that have been localized in this area was *eyeless*. I said nothing and continued to examine the printout. With a slightly triumphant undertone she said, "I told you, it has no homeobox," but I asked her to put the next five hundred base pairs into the computer because I knew that the mouse *Pax-6* gene does have a homeobox. When she did that, a huge list of homeobox genes poured out. Then, when I told her and the people standing around the computer that she had probably cloned *eyeless*, raising the possibility that the *Small eye* gene (=*Pax-6*) of the mouse might be homologous to *eyeless* in *Drosophila*, there was tremendous excitement and she went down the hall hopping from one foot to the other in great joy.

This discovery was quite accidental, because for Quiring's thesis project I had asked her to clone another gene whose protein product we expected to bind to an oligonucleotide sequence that was found upstream of both *fushi tarazu* and *caudal*. As a control I gave her a homeodomain binding site to which we were expecting many homeodomain proteins to bind. She never succeeded in cloning that gene, however, and I finally agreed that she should characterize a clone whose product bound tightly to the control oligonucleotide. Sometimes the control experiments are more important than the

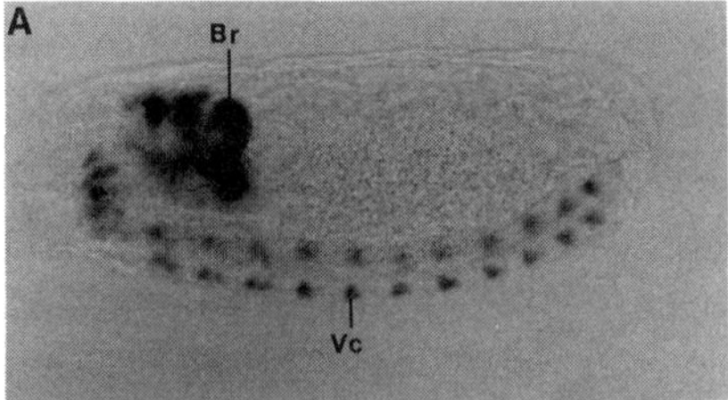

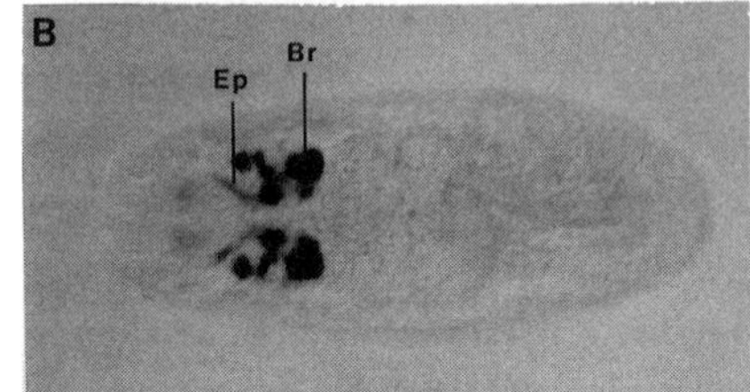

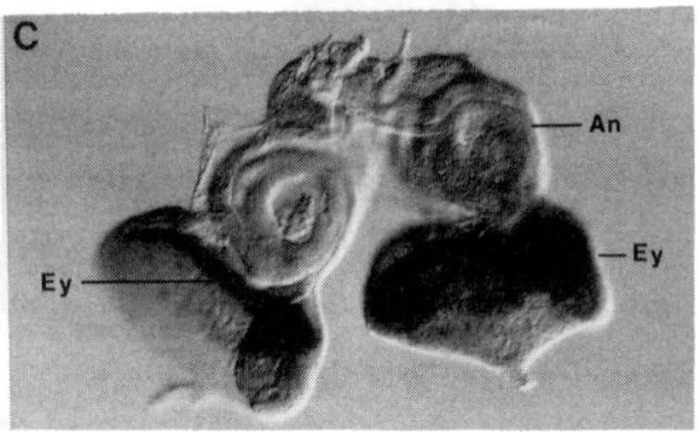

Figure 13.2
Expression pattern of eyeless *in the* Drosophila *embryo and the eye imaginal discs of the larva. (A) Lateral view of the embryo. Expression of* eyeless *transcripts are detected in specific cells of the brain (Br) and ventral chord (Vc). (B) A dorsal view reveals a specific pattern in the brain and the embryonic eye primordia (Ep). (C)* Eyeless *messenger RNA is detected most strongly in the eye imaginal disc, anterior to the morphogenetic furrow. Courtesy of U. Kloter.*

real experiments, and this was true for *eyeless*. Quiring had stimulated my curiosity for the gene that bound to the control oligonucleotide by claiming that even though it bound to a homeodomain binding site, it did not have a homeobox, which later turned out to be false. In any case, this clone was an extremely lucky choice.

With the substantial help of Uwe Walldorf, the question of whether the cloned gene was indeed *eyeless* was tackled. The first *eyeless* mutation had been found by Mildred Hoge as early as 1915 (fig. 13.1). Various *eyeless* mutations collected since then were kept at the *Drosophila* stock centers, but some of the stocks became mixed up, and only three or four mutations could be recovered. However, two of these mutations, which arose spontaneously, were found to be associated with transposon insertions, as is frequently the case for spontaneous mutations in *Drosophila*. This allowed us to show that at least two independent *eyeless* mutations were associated with molecular lesions in the cloned gene, suggesting that the cloned gene represents *eyeless*. This conclusion was strongly reinforced by the study of the expression pattern: the cloned gene is expressed specifically in the eye imaginal discs of normal larvae (fig. 13.2), whereas little or no expression is detectable in the eye discs of *eyeless* mutants. These findings indicated that Quiring indeed had cloned *eyeless*, and the computer search revealed the extensive sequence homology to the murine *Small eye* and the human *Aniridia* genes.

At this point I should like to mention that soon after we had discovered the homeobox in vertebrates, Brigid Hogan had written a letter to me in which she proposed that the murine *Small eye* mutation might be a good candidate for a mutation in a vertebrate homeobox gene. Mice heterozygous for *Small eye* (with one normal and one mutated copy of the gene) have severely reduced eyes, and homozygous embryos (with two defective copies of the gene) have no eyes at all and lack nasal openings. At that time Hogan's arguments did not seem particularly convincing to me, but her intuition was certainly right, because a few years later, in collaboration with Robert Hill, she showed that *Small eye* mutations do indeed affect the *Pax-6* gene, which has a homeobox as well as a paired box.

The paired box-containing genes, the *Pax* genes, were first discovered in *Drosophila* by Markus Noll at the Biozentrum in Basel, and by using the *Drosophila* probes (provided by Herbert Jäckle), Peter Gruss and his collaborators systematically isolated their mouse homologues. The paired box encodes a 130-amino-acid DNA-binding domain that is characteristic for the *Pax* genes. In addition to the paired box, some *Pax* genes contain a homeobox. The *Pax* genes illustrate particularly well what François Jacob has called evolutionary tinkering. The various *Pax* genes are generated by assembling bits and pieces (fig 13.3): some have a paired box only, some have both a paired box and a homeobox, and some have a paired and a partial homeobox encoding the N-terminal arm and the first α-helix only. It therefore appears that new genes are generated either by duplication of entire genes and subsequent divergence, as discussed in the last chapter, or by tinkering—recombination of exons or other DNA segments from different sources—but hardly ever by starting from scratch. When comparing the *Pax* genes of *Drosophila* with those of mammals, Noll noticed that a gene corresponding to *Pax-6* was missing in *Drosophila,* and this was the gene we subsequently discovered. No one would have predicted, however, that the *Pax-6* homologue in *Drosophila* is *eyeless.*

The comparison between *Pax-6* and *eyeless* reveals an unexpectedly high similarity of their sequences: the amino acid sequence of the paired domains

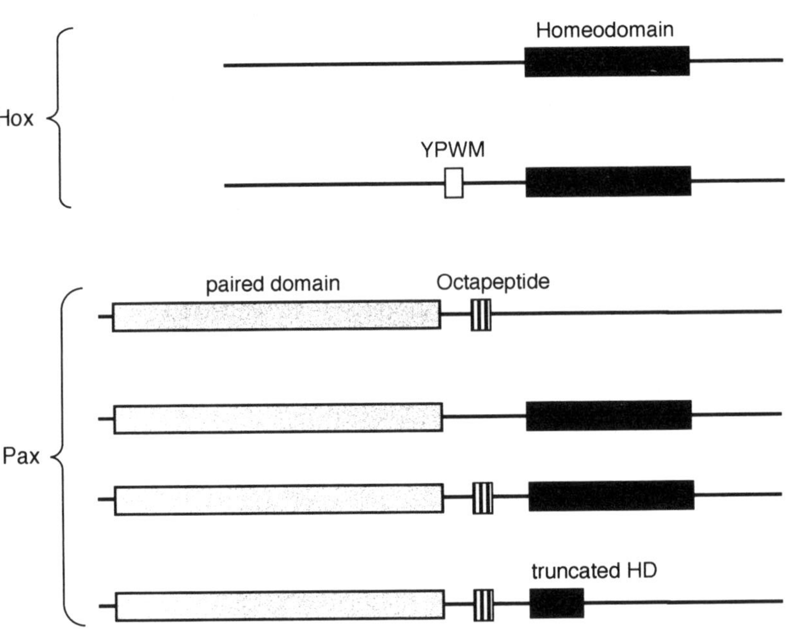

Figure 13.3
"Evolutionary tinkering" with Hox *and* Pax *genes. Differing combinations of homeo- and paired-domains with YPWM and octapeptide motifs are found in various* Hox *and* Pax *genes.*

is 94 percent identical, that of their homeodomains 90 percent, and there are some sequence similarities in the linker between the two boxes. In addition to this high degree of sequence conservation, two of the three intron splice sites in the paired box and one of the two splice sites in the homeobox are conserved, providing a strong argument that *eyeless* and *Pax-6* are true homologues (orthologues).

A comparison of the expression patterns of the two genes is also quite interesting. *Pax-6* is expressed first in the forebrain and hind brain of the mouse embryo and then in the neural tube (the future spinal cord) all along the antero-posterior axis. From the beginning of eye development, *Pax-6* is expressed consecutively in the optic cup, the retina, the lens, and the cornea, suggesting that this gene is involved in the induction of the eye. It is also expressed in the nose. In humans a heritable disease called aniridia has been described that is very similar to the mouse *Small eye* mutation. Heterozygous aniridia patients have severely reduced eyes, with a reduction or complete absence of the iris. Only one lethal fetus has been described that presumably was homozygous for the *Aniridia* mutation, and it lacked eyes completely. Carl Ton, Grady Saunders, Veronica van Heyningen, and their collaborators have cloned the human *Aniridia* gene and found it to be extremely similar to the *Small eye* gene of the mouse; surprisingly, the human and murine proteins have the identical amino acid sequence. The expression patterns in the eye are also essentially the same. The expression of *eyeless* in *Drosophila* is quite similar, beginning with the expression in the brain and along the entire ventral nervous cord of the embryo (see fig. 13.2). Later it becomes confined to certain areas of the brain and the embryonic primordia of the eye. During larval stages, *eyeless* is expressed in the eye discs, most strongly in those cells anterior to the morphogenetic furrow, before their differentiation into the various cell types that make up the eye. This is compatible with an early determinative role of the gene in eye morphogenesis. Comparison of the two expression patterns therefore suggests that in both mammals and insects this gene plays an important role in inducing or determining eye morphogenesis. In addition, the fact that the two genes are expressed in the dorsal spinal cord and the ventral nerve cord, respectively, further supports the idea that the body plans of vertebrates and invertebrates are dorso-ventrally inverted relative to each other.

Are *eyeless* and *Small eye* master control genes of eye morphogenesis? A priori *eyeless* did not seem to be a switch gene, because its (partial) loss-of-function mutations lead to a reduction of the eyes rather than to their homeotic transformation. I therefore wanted to construct a gain-of-function mutant, hoping that the ectopic expression of the *eyeless* gene might induce the formation of eyes in other regions of the body. This experiment met with deep skepticism among my colleagues, however. I reported the cloning of the *eyeless* gene and its homology to *Pax-6* at our *Drosophila* Workshop in Kolymbari on the Greek island of Crete. Every two years the Drosophilists from all over the world meet in this beautiful place for intensive discussions of their latest scientific results. After my presentation of the astonishing fact that homologous genes are involved in eye morphogenesis of both insects and vertebrates, Rolf Bodmer presented data that the homeobox gene *tinman,* that he is working on, seems to be involved in the development of the heart in both insects and mammals. Because insects have a dorsal heart and vertebrates have a ventral one, this is another argument in favor of Saint-Hilaire's hypothesis of the inversion of the body plans. There followed a vivid discussion about my dream experiment of inducing additional ectopic eyes by the expression of the *eyeless* gene in imaginal discs other than the eye disc, for example, in wing discs. Each of my colleagues found at least one good reason why this crazy experiment could not possibly work, and the discussion even continued in the afternoon when we went for a swim in the sea. I told my younger colleagues that I had observed transdetermination of wing to eye tissue (see fig. 11.4) some twenty years ago, but they were convinced that a single gene could not possibly induce an eye.

Back in the laboratory I convinced Patrick Callaerts, a postdoctoral fellow from Belgium, to try my dream experiment, not only with *Drosophila* but also with the *Small eye* gene of the mouse. Callaerts was later reinforced by Georg Halder, a graduate student, who had some experience with targeted gene expression. We decided to make use of the yeast transcription factor GAL4 for directing the expression of *eyeless* into imaginal discs other than the eye disc where it is normally expressed (fig. 13.4). Andrea Brand and Norbert Perrimon had taken our enhancer detection method a step further and used the yeast transcription factor GAL4 as a reporter gene rather than using β-galactosidase. GAL4 has no homologue and no known target genes in *Drosophila,* making it useful for targeting gene expression. The gene to be expressed, in our case *eyeless,* is fused to a regulatory element (UAS) to which GAL4 can bind and activate its expression. For this purpose we first had to isolate enhancer detector lines expressing GAL4 in various imaginal discs. Line E132 expresses GAL4 in the antennal, wing, and leg discs (fig. 13.4) and therefore allows the targeted expression of *eyeless* in those discs.

As in the case of *Antennapedia,* where at first only three bristles were detected, Callaerts and Halder initially found only some red eye pigment, but a week later the first eye facets appeared and then spectacular ectopic eyes were revealed on antennae, wings, and legs (pl. 7, fig. 13.5). The analysis of the ectopic eyes by light and scanning electron microscopy revealed the presence of morphologically normal eye structures with normal photoreceptor cells (fig. 13.6). By shining light on the ectopic eye the photoreceptors were shown to produce an electrical response that is normal for isolated photoreceptor cells. These findings show that *eyeless* is a master control gene for eye morphogenesis. A single gene can turn on a cascade of some 2,500 genes that are required to "build" an eye. This value is based on large

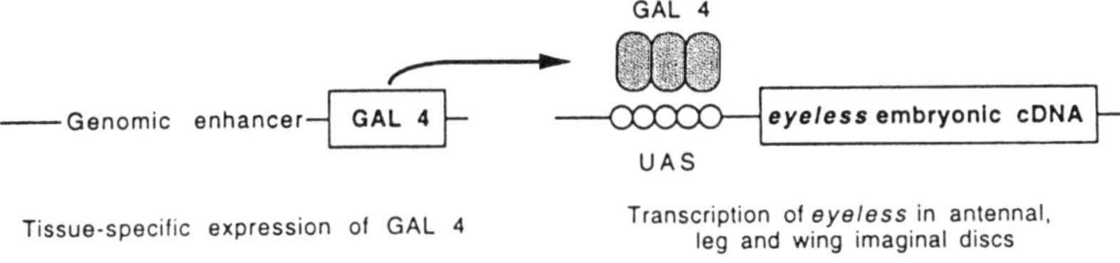

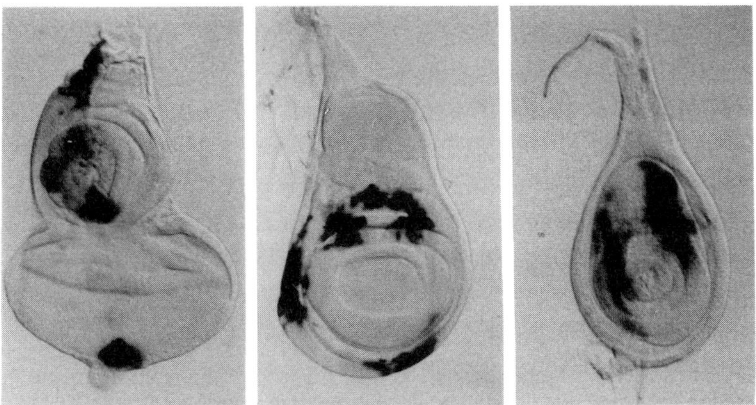

Figure 13.4
Targeted expression of the eyeless *gene. The yeast transcription factor GAL4 is tissue-specifically expressed under the control of a genomic enhancer. Its expression patterns are shown (from left to right) in the eye-antennal disc, the wing disc, and the leg disc. The* eyeless *gene (embryonic cDNA) is fused to five repeats of the upstream activating sequence (UAS) through which GAL4 drives the expression of* eyeless. eyeless *expression is confined to those areas where GAL4 is expressed. From G. Halder, P. Callaerts, and W. J. Gehring, Induction of ectopic eyes by targeted expression of the* eyeless *gene in* Drosophila, Science *267 (1995): 1788–92, with permission from the American Association for the Advancement of Science (AAAS), Washington, D.C.*

enhancer detection screens carried out in Gerry Rubin's laboratory. The induction of ectopic eyes by a single gene perhaps illustrates the term *master control gene* best.

By this time, everybody in the laboratory was convinced that my predictions were correct, and so it came as less of a surprise when the mouse gene was also shown to be capable of inducing eyes (fig. 13.7). Of course, these eyes were *Drosophila* eyes, as expected on the basis of Oscar Schotté's experiment (see Chapter 4), because the mouse provided only the switch gene and *Drosophila* contributed the other 2,500 genes required to make an eye.

This time our findings appeared on the front page of the *New York Times* under the title "With New Fly, Science Outdoes Hollywood," and the public response was much stronger than when we had induced antennal legs. The eye is a very special organ.

The identification of *eyeless* as a master control gene for eye development has strong repercussions on our thinking about eye evolution. Already for Darwin, the evolution of eyes was a difficult problem. In *The Origin of*

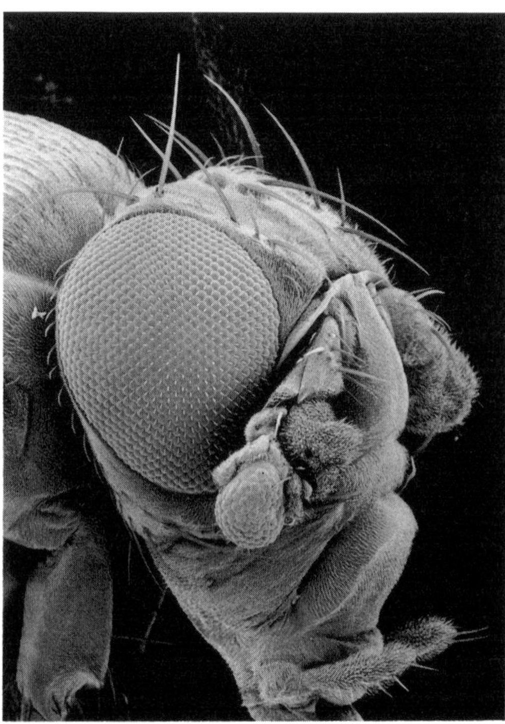

Figure 13.5
Induction of an additional eye on the antenna of a fruit fly by ectopic expression of the eyeless *gene. Scanning electron micrograph by G. Halder and A. Hefti.*

Species he devoted a chapter to "Difficulties of the Theory" and mentioned the difficulty of explaining the origin of the seemingly perfect eye. A priori he argued that one would think it impossible that such a perfect organ as the eye could evolve by random variation and selection. (Instead of random variation, we would say mutations; Darwin did not know the basis of genetics.) However, he continued, if the perfect eyes that we see in living organisms were derived from a primitive prototype eye consisting of a light-sensitive nerve cell and a pigment cell, selection might explain the evolution of increasingly better eyes that would confer a selective advantage to the respective organism. The three major animal phyla have evolved different solutions to the problem of obtaining an image of the outside world (pl. 8): the eyes of vertebrates and cephalopods (mollusks) contain a single lens and are built similarly to a photocamera, whereas arthropods (including insects) have compound eyes consisting of multiple facets each containing its own lens (fig. 13.8). Traditionally, these eye types have been considered to be of independent evolutionary origin. On one hand, the morphology of the com-

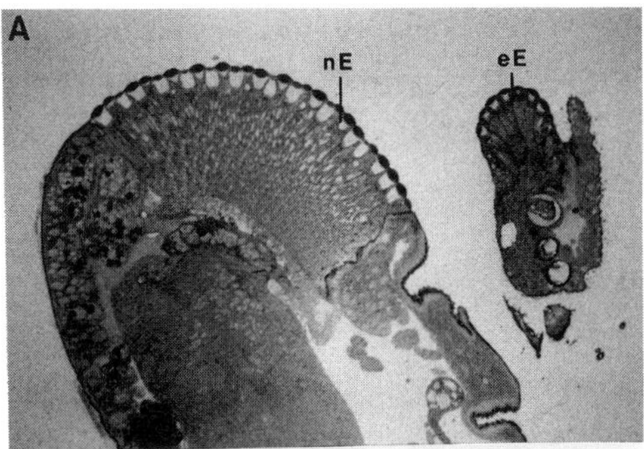

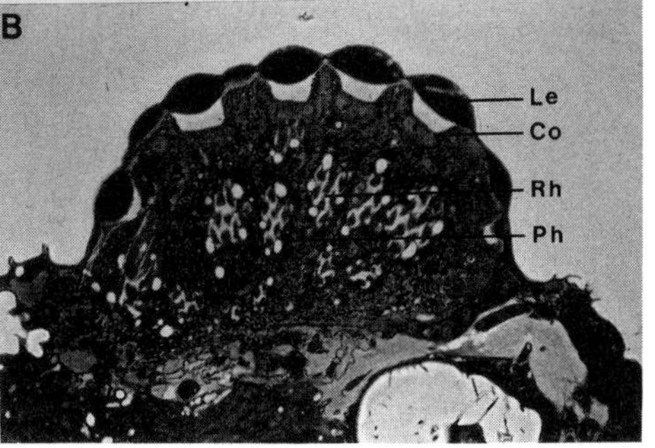

Figure 13.6
Morphology of a normal and an ectopic eye: (A) section across a normal eye (nE) and an ectopic eye (eE) on the antenna; (B) normal morphology of the ectopic eye with lenses (Le), cone cells (Co), photoreceptors (Ph), and the normal number of rhabdomeres (Rh). From G. Halder, P. Callaerts, and W. J. Gehring, Induction of ectopic eyes by targeted expression of the eyeless *gene in* Drosophila, Science 267 *(1995): 1788–92, with permission from the American Association for the Advancement of Science (AAAS), Washington, D.C.*

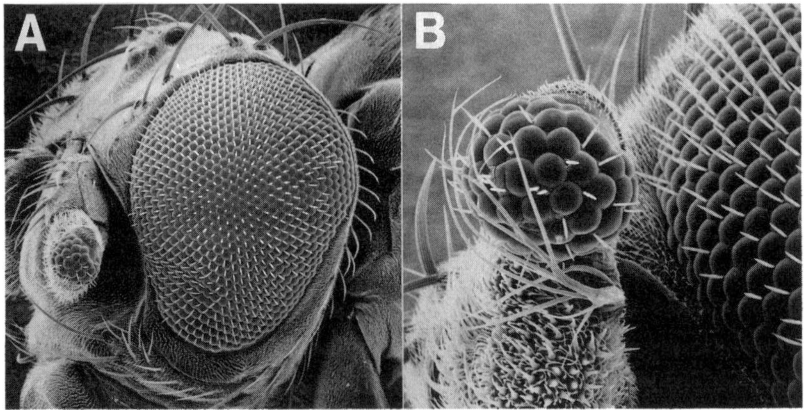

Figure 13.7
Eye induced on the antenna of a fly with the mouse Pax-6 (Small eye) *gene: (A) overview; (B) higher magnification. Scanning electron micrograph by G. Halder and A. Hefti.*

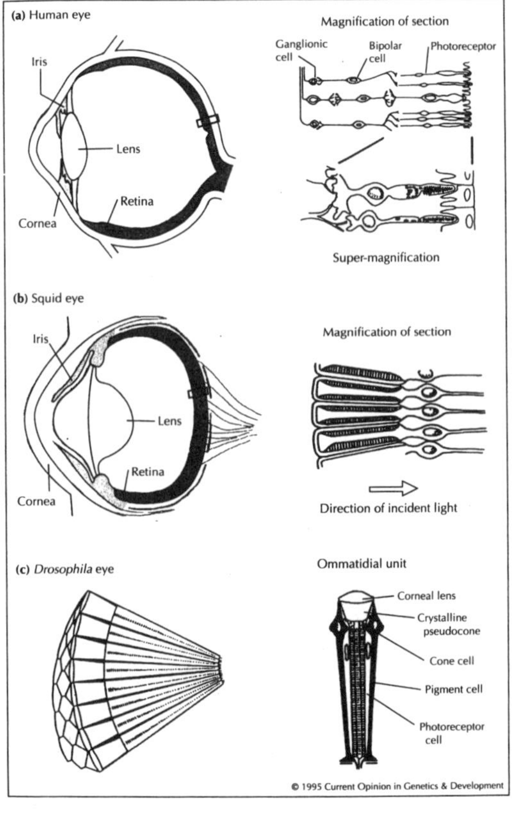

Figure 13.8
Three major eye types: (a) human eye—the optic cup is formed as an evagination of the brain that leads to an inversion of the photoreceptor cells, which point away from the incident light; (b) squid eye—the optic cup is formed as an invagination of the skin, and the photoreceptor cells are oriented toward the incident light; (c) Drosophila *eye—the compound eye of insects consists of numerous ommatidial units and has a different morphology and mode of development than the single lens camera–type eyes of vertebrates and cephalopods. From G. Halder, P. Callaerts, and W. J. Gehring, New perspectives on eye evolution,* Current Opinion in Genetics and Development 5 *(1995): 602–9.*

pound eye of insects and the single lens eye of vertebrates and mollusks are so strikingly different that no structural homology can be claimed. On the other, the mode of development of the cephalopod eye and the vertebrate eye implies that these two systems evolved independently. The cephalopod eye develops from an invagination of the skin and leads to an everted arrangement of the photoreceptor cells, whereas the vertebrate eye forms as an evagination of the brain leading to an inverted arrangement of the photoreceptors. Based on these substantial differences in morphology and mode development the biologist Ernst Mayr has argued that different types of eyes evolved as many as forty times independently in the animal kingdom. Because the evolution of the prototype eye, at a stage before selection can exert its effect, must be a rare event, the independent evolution of so many prototypes represents a serious problem that is difficult to reconcile with Darwin's theory.

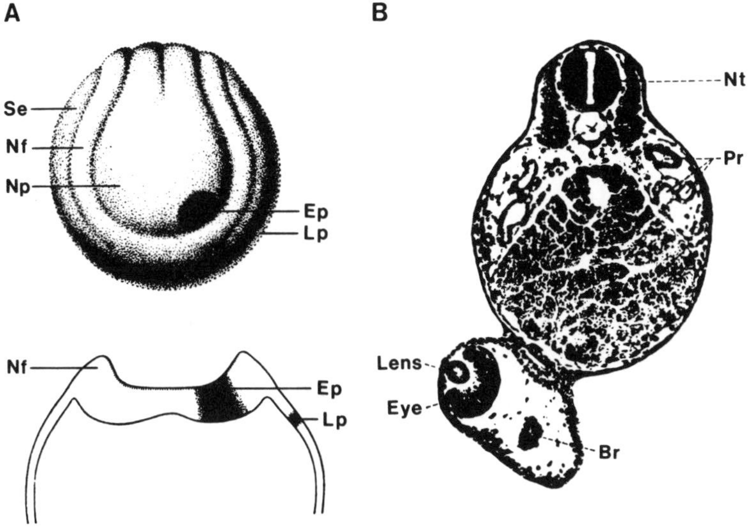

Figure 13.9
Induction of an ectopic eye with lens by transplantation of the eye primordium in a newt embryo. (A) Localization of the eye primordium (optic cup) (Ep) and the lens primordium (Lp) at the neurula stage; Nf *neural fold,* Np *neural plate,* Se *skin ectoderm. (B) Upon transplantation of the eye primordium (Ep) underneath the skin ectoderm into the flank of a recipient embryo, an ectopic eye (Eye) with a lens induced in the skin ectoderm of the recipient develops;* Br *brain structures,* Pr *pronephric ducts,* Nt *neural tube (cross-section). From H. Spemann,* Embryonic Development and Induction *(New Haven: Yale University Press, 1938).*

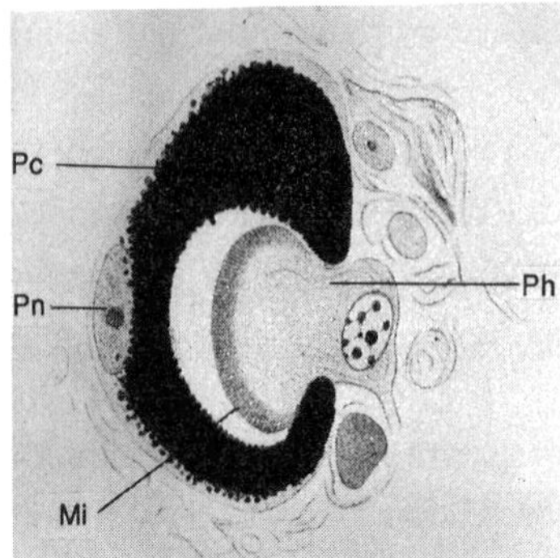

Figure 13.10
Primitive eye of the flatworm Planaria torva. *This eye, which consists of a single pigment cell (Pc) with its nucleus (Pn) and three photoreceptor cells (Ph) (only one is shown in this section) with microvilli (Mi), closely resembles the prototypic eye postulated by Darwin. From R. Hesse, in* Zeitschrift für wissenschaftliche Zoologie *62 (1990): 5–246, pl. 27, fig. 2.*

The mechanism of eye development was first studied experimentally in amphibia by Hans Spemann, who localized the primordia for the optic cups in the anterior neural plate, in the area of the forebrain (fig. 13.9), whereas the primordia for the future lens was located outside of the neural plate in the ectoderm of the skin. By transplanting the primordia of the optic cup to an ectopic position on the ventral side of the embryo, Spemann obtained the development of an ectopic eye and induction of a lens in the overlaying skin ectoderm. This was the first induction of ectopic eyes, but Spemann did not consider gene action in explaining the mechanism of induction.

The genetic cascade controlling eye morphogenesis in *Drosophila* has been analyzed by using the *sevenless* gene as an entry point by the combined efforts of many excellent laboratories, such as those of Seymour Benzer, Gerry Rubin, Don Ready, Ernst Hafen, Larry Zipursky, and Konrad Basler. Even though some of the genes involved in the signal transduction pathway are found in both mammals and *Drosophila,* there did not seem much in common, although the basic components of the visual process, like the photoreceptor pigment rhodopsin are shared between insects and mammals. The upper parts of this genetic cascade had not been elucidated, however, and that is where *eyeless* plays its fundamental role.

The *eyeless* master control gene of *Drosophila* and mouse can induce eye morphogenesis. Since this discovery we have cloned the *eyeless* homologue of the squid, in collaboration with Joram Piatigorsky and his collaborators, and have shown that it, too, is capable of inducing eyes when expressed ectopically in *Drosophila.* Because we also have cloned highly conserved *Pax-6* homologues from sea squirts, nemertine worms, and flat worms and have shown that *Pax-6* is expressed specifically in the eyes, we consider *eyeless* (*Pax-6*) to be a universal master control gene for eye morphogenesis. These findings lead to the further conclusions that the prototypic eye may have originated only once, rather than some forty times, and that the large variety of eye types found in the animal kingdom is derived from this prototype by divergent, parallel, and convergent evolution. In fact, the eye of certain flatworms resembles the Darwinian prototype very closely (fig. 13.10), and in one species found in Japan, the eyes consist of a single photoreceptor and a single pigment cell, as Darwin postulated. I presume that Charles Darwin would have been pleased with this conclusion, but a lot more work needs to be done to prove this revolutionary hypothesis. I do think, however, that we now have the necessary tools to begin to study these evolutionary problems experimentally. In any case, the idea of the magnificent eye has been taken up at the annual Carneval in Basel (fig. 13.11) and was widely accepted.

In this book I have tried to give the reader a guided tour through the homeobox story leading from antennal legs to the DNA sequence of the homeobox and to the structure of the homeodomain at atomic resolution, from the living flies looking back in time to the origins of evolution of the body plan and of the eyes. I hope to have conveyed to my readers some of the excitement that is generated by the biological research on the little fruit fly called *Drosophila.*

Figure 13.11
The "magnificent eyes" at the Carneval in Basel. Courtesy of G. Halder.

Appendix 1

Theories About Fertilization
Letter of F. Miescher to W. His
Basel, December 17, 1982

With predilection I am reading literature on plant biology. There one finds the most fundamental and general viewpoints on sexuality. (Most profoundly by Darwin on cross- and hybrid fertilization in plants.) All phenomena of animal sexuality are already full of specific adaptations, which obscure the general principle. For me, the key to sexuality lies in the stereochemistry. The "gemmules" of Darwin's pangenesis are nothing else than the numerous asymmetric carbon atoms in the organized substances. These carbon atoms through minute causes and changes in external conditions undergo conformational changes, whereby gradually mistakes arise in their organization. Sexuality provides a mechanism to correct these unavoidable stereometric architectural errors in the structure of organized substances. Left-handedness is corrected by right-handedness, and the equilibrium is restored. In the enormous protein molecules (egg white bodies) or in the even more complex molecules of hemoglobin, etc., *the numerous asymmetric carbon atoms allow a colossal amount of stereoisomers, so that all the richness and all variations of hereditary transmissions can find their expression equally well, as the words and terms of all languages in the 24–30 letters of the alphabet.* Therefore, it is superfluous to propose that egg- and sperm cells, or cells in general, are storage cupboards for innumerable chemical substances, each one of which is supposed to be a carrier of a special heritable character (de Vries pangenesis). Protoplasm and nucleus, judging from my own studies, do not consist of innumerable chemical substances, but rather of a very few individual chemicals, however, they are probably of very complicated structure.

Appendix 2

Sperm and Inheritance
Letter of F. Miescher to W. His
Basel, October 13, 1893

Now I also collect material from trout sperm (in addition to salmon sperm, ed.). For the theory of inheritance it is interesting, whether one can find small but already significant chemical differences on the sperm heads of two so closely related animal species. In doing this, I encounter the fact that chemists have little sense for such small differences, and even less recognition signs. Physical chemistry provides at least some cues.

My spermatogenesis campaign was strongly impaired by the low yield in catching fish during the most important period (from September 15 to 30). Nevertheless, I have collected some material and it will become apparent what the analyses will give. Probably there is not a single substance in the spermatocytes identical with what is later found in sperm. With morphological continuity it is not so easy to explain how out of three substances four should arise and phosphorus, sulfur, iron and nitrogen become distributed in the most bizarre way. *The continuity* lies not only in the form, it also lies deeper than the chemical molecule. *It lies in the groups of atoms that constitute the molecule. In this sense I am a proponent of the chemical theory of inheritance . . . à outrance.* But one has to keep in mind that the properties of chemical compounds are based on the nature and intensity of the movements of the atoms, and that the intramolecular movements of the atoms in these easily degradable biological substances have an especially large intensity and independence when compared to the inertia of the whole molecule, this is why they are degradable.

The speculations of Weissmann et al. torment themselves with pseudochemical terms, which are partially unclear and partially correspond to an outdated status of chemistry. *If, as it is easily possible, a protein molecule contains 40 asymmetric carbon atoms, this results in* 2^{40}, i.e. approximately a trillion ($=10^{12}$) isomers. And this is only one kind of isomer, whereas the isomerism of nitrogen and the unsaturated valencies are not considered. *In order to account for the immense variability postulated by the theory of inheritance, my theory is better suited than any other.* In this connection all transitions from the hardly recognizable to the largest differences can be thought of, which, however, requires a very critical discussion of this question.

Appendix 3

Drosophila Life Cycle

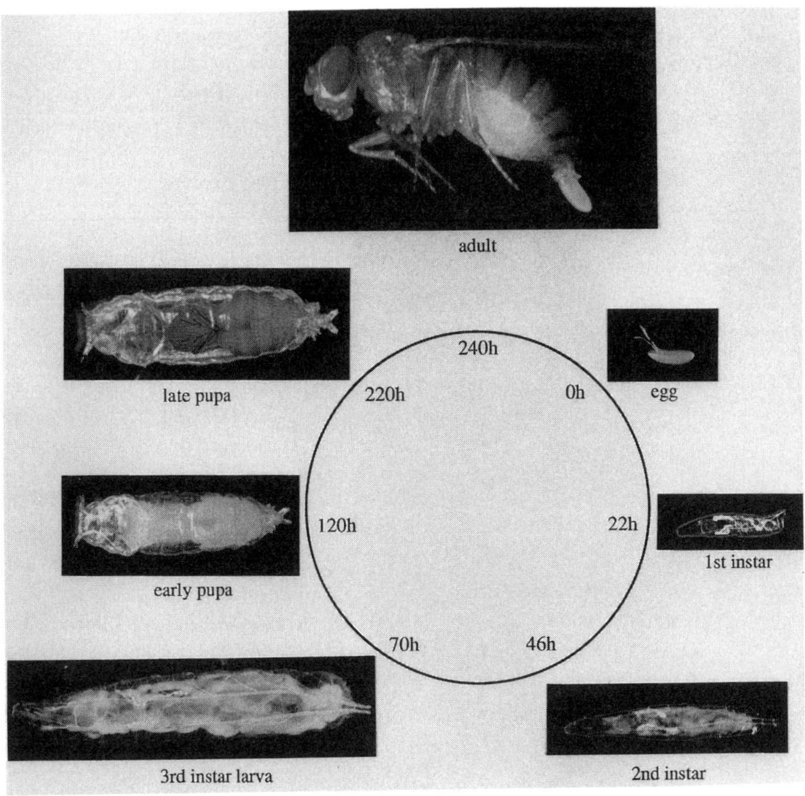

Courtesy of U. Kloter.

Glossary

activator transcriptional protein that binds to a specific region of DNA and stimulates transcription of a nearby gene

allele one of a series of possible alternative variants of a given gene

alpha helix (α-helix) common structural motif in proteins in which a linear sequence of amino acids folds into a right-handed helix stabilized by internal hydrogen bonds

antibody immunoglobulin proteins produced by lymphocytes that bind tightly to a foreign antigen molecule

antibody staining detection of antigenic molecules that bind to a specific antibody molecule by labeling, e.g., with fluorescent dye

antigen molecule that provokes an immune response, e.g., synthesis of a specific antibody

apoptosis programmed cell death

archenteron primary gut formed during gastrulation

autoregulation regulation of the synthesis of a gene product by the gene product itself

bacteriophage virus that infects bacteria (from Greek *phagein,* to eat)

base molecule that accepts proton in solution; the bases (A adenin, G guanin, C cytosin, T thymin, U uridine) constitute the building blocks of nucleic acids; in DNA, the bases are either purines (A or G) or pyrimidines (C or T); in RNA, T is replaced by U

base pair two bases paired by hydrogen bonds, e.g., G with C or A with T

blastocyst thin-walled hollow sphere formed early in mammalian development consisting of an outer wall, the trophoblast, and the inner cell mass representing the embryo proper

blastoderm in eggs containing a large amount of yolk the cleavage (early cell divisions) is restricted to a superficial layer of the fertilized egg; the stage when this layer forms is termed the blastoderm and corresponds to the blastula stage of animals with complete cleavage

blastomere embryonic cell formed by cleavage divisions of the egg

blastopore site where the cells on the surface of the embryo move inside the embryo to form the inner germ layers (mesoderm and endoderm) at the gastrula stage

blastula stage of embryonic development near the end of cleavage; usually a hollow ball of cells

chromosome threadlike structure in the cell nucleus consisting of DNA in which the genes are linearly arranged, associated with proteins and RNA

cleavage early cell divisions of the fertilized egg

clone family of cells derived from a single mother cell by repeated division

cloning production of multiple copies of a gene by repeated replication

cloning vector genetic element, usually a plasmid or a bacteriophage, used as a carrier to transfer a piece of DNA into a recipient cell for the purpose of gene cloning

codon sequence of three nucleotides in a DNA or messenger RNA molecule that represents the instruction for incorporation of a specific amino acid into protein

colinearity rule homeotic genes are generally arranged in clusters and arranged along the chromosome in same order as they are expressed along the antero-posterior axis of the embryo

complementation production of the normal phenotype when two different mutant alleles are combined in a heterozygous diploid

cytoplasm substance contained within the plasma membrane excluding the nucleus in eukaroytic cells

determination commitment of a cell or embryonic tissue to a certain developmental pathway

diploid cell or organism having two sets of chromosomes

dissociation constant measure of the tendency of a complex of molecules to dissociate

DNA deoxyribonucleic acid; carrier of genetic information; single- or double-stranded polynucleotide chain consisting of bases or base pairs, respectively, linked to a backbone of sugar (deoxyribose) and phosphate groups; double-stranded DNA molecule forms a double helix

dominant mutation that manifests its effect in heterozygous condition

ectoderm outer germ layer of the embryo

endoderm inner germ layer of the embryo

enhancer DNA sequences that potentiate the transcriptional activity of physically linked genes

enzyme protein that catalyzes a specific chemical reaction

feedback the influence of the result of a process on the process itself

fertilization fusion of the haploid egg and sperm to form a diploid cell, the zygote

gap genes genes expressed in broad domains of the early embryo comprising several body segments; loss-of-function mutations lead to the corresponding gaps in the pattern

gastrula stage of embryonic development when cellular movements lead to the formation of the three primary germ layers

gene fundamental functional unit of heredity that carries the information from generation to generation; consists of a segment of DNA composed of a transcribed region and regulatory sequences

genome genetic material of an organism

gynandromorphs genetic mosaic animals consisting of both male and female cells

haploid cell or organism having one set of chromosomes

heterotopic transplantation transplantation of a cell or tissue to a different site of another embryo or animal

heterozygous two different alleles present at a given genetic locus

homeobox conserved DNA segment of 180 base pairs shared by homeotic genes

homeodomain DNA binding protein domain of sixty amino acids encoded by the homeobox

homeosis change of something into the likeness of something else (Bateson); transformation of one body segment or part thereof, into the corresponding structures of another body segment

homeotic genes genes that specify the identity and sequence of the body segments

homeotic mutation mutation transforming one part of the body into another

homologous referring to genes, structures, or processes in different organisms that show fundamental similarity because of their descendance from a common ancestor

homologous recombination genetic recombination based upon DNA sequence homology

homotopic transplantation transplantation of a cell or tissue to the same site of another embryo or animal

homozygous two identical alleles present at a given genetic locus

hybridization nucleic acid hybridization is a detection method for specific nucleotide sequences; a process whereby two complementary nucleic acid strands form a double helix

imaginal disc disc of cells in insect larvae that forms adult structures (e.g., legs, wings) during morphogenesis

in situ hybridization technique in which a single-stranded DNA or RNA probe is used to locate a gene or messenger RNA molecule in a cell, tissue, or complete organism

intron intervening sequence in split genes that is transcribed into RNA but is subsequently removed from mRNA by splicing

inversion chromosome mutation that reverses the gene order between two chromosomal breakpoints

macromere large blastomere

maternal coordinate genes genes expressed during oogenesis that specify a coordinate system of positional information in the egg

mesomere medium-sized blastomere

mesoderm middle germ layer of the embryo

messenger RNA (mRNA) RNA molecule that specifies the amino acid sequence of a protein; it is synthesized by RNA polymerase as a complementary copy of one DNA strand and translated into protein in a process catalyzed by ribosomes

microfilament elongated intracellular fibers consisting of polymerized actin molecules involved in cell movement and cell structure

micromere small blastomere

microtubule long, thin cylindrical organelles involved in intracellular transport, cell division, ciliary motion, etc.

mitochondrion self-reproducing organelle in the cytoplasm of most eukaryotic cells that function in respiration and oxidative phosphorylation (i.e., the production of the energy-rich compound adenosine triphosphate)

neural crest group of embryonic cells that emigrate from the roof of the neural tube and give rise to various types of adult cells, including pigment cells, and peripheral nerve cells

neurula stage of embryonic development when the central nervous system begins to form

nuclear magnetic resonance (NMR) resonant absorption of electromagnetic radiation at a specific frequency by atomic nuclei in a magnetic field due to flipping of the orientation of their magnetic dipole moments; employed to determine the three-dimensional structure of small proteins and nucleic acids in solution

nucleic acids DNA or RNA; chains of nucleotides joined together by phosphodiester bonds

nucleoside compound molecule composed of a purine or pyrimidine base linked to either a ribose or a deoxyribose sugar moiety

nucleotide nucleoside with a phosphate group linked to the sugar moiety by an ester linkage

nucleus envelope-bounded organelle in eukaryotic cells containing the chromosomes

organelle any complex structure that forms a component of cells and performs a characteristic function (e.g., nucleus, ribosome, mitochondrion)

operon group of contiguous genes in bacteria that are transcribed as a single messenger RNA molecule

orthologous referring to genes in different species that have originated from a single ancestral gene by a speciation event

pair-rule genes genes expressed in stripes (belts) along the blastoderm embryo with a periodicity that corresponds to every other segment (or parasegment)
paralogous referring to two or more genes in the same species that are so similar in their nucleotide sequences that they are assumed to have originated from a single ancestral gene by gene duplication
placenta organ consisting of embryonic and maternal tissues in close union through which the mammalian embryo is nourished
proliferation growth by cell division
promoter DNA region to which the RNA polymerase binds and initiates transcription of RNA synthesized on the DNA template
protein linear chain of amino acids linked by peptide bonds
recessive mutation masked by the presence of a dominant allele and manifesting its effect only in homozygous condition
recombinant DNA technology technology based on joining DNA fragments from different sources, inserting them into cloning vectors for their cloning in suitable host cells
recombination occurrence of progeny with combinations of genes other than those that occurred in their parents due to independent assortment or crossing-over
replication DNA synthesis; the two DNA strands are separated and each is copied into a complementary strand by an enzyme complex containing DNA polymerase
repressor protein that binds to a specific region of DNA to prevent transcription of a nearby gene
ribosome particle consisting of two subunits composed of ribosomal RNAs and ribosomal proteins that associates with messenger RNA and catalyzes the synthesis of proteins
RNA (ribonucleic acid) single-stranded polynucleotide chain consisting of bases linked to a backbone of sugar (ribose) and phosphate groups
tail bud stage of embryonic development when the basic organization of the embryo is laid down, except for the tail, which still forms a budlike rudiment
segmentation genes genes whose expression specify the pattern of body segments
segment polarity genes genes that regulate the pattern within each segment
transcription copying of one strand of DNA into a complementary RNA sequence by the enzyme RNA polymerase
transformation gene transfer by external application of DNA
translation process occurring on the ribosome by which the sequence of nucleotides in a messenger RNA molecule directs the incorporation of amino acids into protein

trophectoderm extra-embryonic part of the ectoderm of mammalian embryos at the blastocyst stage

trophoblast extra-embryonic cell layer surrounding the embryo proper in mammals and attaching the embryo to the uterus wall

unequal crossing-over exchange of genetic material between chromosomes that are not precisely paired, resulting in nonreciprocal exchange leading to duplication and deletion, respectively, of a DNA segment

Further Reading

Chapter 1

GENERAL

Jacob, F. *The Statue Within: An Autobiography*, trans. F. Philip. Plainview, N.Y.: Cold Spring Harbor Laboratory Press, 1995.

Watson, J. P. *The Double Helix*. New York: Atheneum, 1968.

Chapter 2

GENERAL

Lewis, E. B. A gene complex controlling segmentation in *Drosophila*. *Nature* 276 (1978): 565–70.

Stümpke, H. *The Snouters: Form and Life of the Rhinogrades*. Chicago: University of Chicago Press, 1981.

Chapter 3

GENERAL

Watson, J. D., and J. Tooze. *The DNA Story: A Documentary History of Gene Cloning*. San Francisco: W. H. Freeman, 1981.

SPECIFIC ARTICLES

Bender, W., M. Akam, F. Karch, P. Beachy, M. Peifer, P. Spierer, E. B. Lewis, and D. Hogness. Molecular genetics of the bithorax complex in *Drosophila melanogaster*. *Science* 221 (1983): 23–29.

Carrasco, A. E., W. McGinnis, W. J. Gehring, and E. M. De Robertis. Cloning of an *X. laevis* gene expressed during early embryogenesis that codes for a peptide region homologous to *Drosophila* homeotic genes. *Cell* 37 (1984): 409–14.

Garber, R. L., A. Kuroiwa, and W. J. Gehring. Genomic and cDNA clones of the homeotic locus *Antennapedia* in *Drosophila*. *EMBO Journal* 2 (1983): 2027–36.

Gehring, W. J. Homeo boxes in the study of development. *Science* 236 (1987): 1245–52.

McGinnis, W., R. L. Garber, J. Wirz, A. Kuroiwa, and W. J. Gehring. A homologous protein-coding sequence in *Drosophila* homeotic genes and its conservation in other metazoans. *Cell* 37 (1984): 403–8.

McGinnis, W., C. P. Hart, W. J. Gehring, and F. H. Ruddle. Molecular cloning and chromosome mapping of a mouse DNA sequence homologous to homeotic genes of *Drosophila. Cell* 38 (1984): 675–80.

McGinnis, W., M. S. Levine, E. Hafen, A. Kuroiwa, and W. J. Gehring. A conserved DNA sequence in homeotic genes of the *Drosophila* Antennapedia and bithorax complex. *Nature* 308 (1984): 428–33.

Rubin, G., and A. Spradling. Genetic transformation of *Drosophila* with transposable element vectors. *Science* 218 (1982): 348–53.

Scott, M., and A. Weiner. (1984). Structural relationships among genes that control development: Sequence homology between *Antennapedia, Ultrabithorax,* and *fushi tarazu* loci of *Drosophila. Proceedings of the National Academy of Sciences* 81 (1984): 4115–19.

Shepherd, J. C. W., W. McGinnis, A. E. Carrasco, E. M. De Robertis, and W. J. Gehring. Fly and frog homoeo domains show homologies with yeast mating type regulatory proteins. *Nature* 310 (1984): 70–71.

Chapter 4

GENERAL

Bate, M., and A. Martinez Arias. *The Development of "Drosophila melanogaster."* Plainview, N.Y.: Cold Spring Harbor Laboratory Press, 1993. Vols. 1 and 2.

Davidson, E. H. *Gene Activity in Early Development,* 3d ed. Orlando, Fla.: Academic Press, 1986.

Gilbert, S. *Developmental Biology,* 5th ed. Sunderland, Mass.: Sinauer Associates, 1997.

Hamburger, V. *The Heritage of Experimental Embryology: Hans Spemann and the Organizer.* New York: Oxford University Press, 1988.

Riddle, D. L., T. Blumenthal, B. Meyer, and J. Priess, eds. *C. elegans II.* Cold Spring Harbor Laboratory Press, 1997.

Satoh, N. *Developmental Biology of Ascidians.* Cambridge: Cambridge University Press, 1994.

Spemann, H. *Embryonic Development and Induction.* New Haven: Yale University Press, 1938.

Wood, W. B., et al. *The Nematode "Caenorhabditis elegans."* Cold Spring Harbor, N.Y.: Cold Spring Harbor Laboratory, 1988.

Chapter 5

GENERAL

St. Johnson, D., and C. Nüsslein-Volhard. The origin of pattern and polarity in the *Drosophila* embryo. *Cell* 68 (1992): 201–19.

Steward, R., and S. Govind. Dorsal-ventral polarity in the *Drosophila* embryo. *Current Biology* 3 (1993): 556–61.

SPECIFIC ARTICLE

Mlodzik, M., and W. J. Gehring. Expression of the *caudal* gene in the germ line of *Drosophila:* Formation of an RNA and protein gradient during early embryogenesis. *Cell* 48 (1987): 465–78.

Chapter 6

GENERAL

Nüsslein-Vollhard, C., and E. Wieschaus. Mutations affecting segment number and polarity in *Drosophila. Nature* 287 (1980): 795–801.

SPECIFIC ARTICLES

Hafen, E., A. Kuroiwa, and W. J. Gehring. Spatial distribution of transcripts from the segmentation gene *fushi tarazu* during *Drosophila* embryonic development. *Cell* 37 (1984): 833–41.

Hiromi, Y., A. Kuroiwa, and W. J. Gehring. Control elements of the *Drosophila* segmentation gene *fushi tarazu. Cell* 43 (1985): 603–13.

Hiromi, Y., and W. J. Gehring. Regulation and function of the *Drosophila* gene *fushi tarazu. Cell* 50 (1987): 963–74.

Rivera-Pomar, R., and H. Jäckle (1996). From gradients to stripes in *Drosophila:* Filling in the gaps. *Trends in Genetics* 12 (1996): 478–83.

Small, S., A. Blair, and M. Levine. Regulation of two pair-rule stripes by a single enhancer in the *Drosophila* embryo. *Developmental Biology* 175 (1996): 314–24.

Chapter 7

GENERAL

Lewis, E. B. Clusters of master control genes regulate the development of higher organisms. *Journal of the American Medical Association* 267 (1992): 1524–31.

SPECIFIC ARTICLES

Hafen, E., M. Levine, and W. J. Gehring. Regulation of *Antennapedia* transcript distribution by the bithorax complex in *Drosophila. Nature* 307 (1984): 287–89.

Levin, M., E. Hafen, R. L. Garber, and W. J. Gehring. Spatial distribution of *Antennapedia* transcripts during *Drosophila* development. *EMBO Journal* 2 (1983): 2037–46.

Schneuwly, S., R. Klemenz, and W. J. Gehring. Redesigning the body plan of *Drosophila* by ectopic expression of the homoeotic gene *Antennapedia. Nature* 325 (1987): 816–18.

Chapter 8

GENERAL

Perutz, M. F. *Protein Structure: New Approaches to Disease and Therapy.* New York: W. H. Freeman, 1992.

Ptashne, M. *A Genetic Switch: Phage "Lambda" and Higher Organisms,* 2d ed. Cambridge, Mass.: Cell Press and Blackwell Scientific, 1992.

Wüthrich, K. *NMR of Proteins and Nucleic Acids.* New York: Wiley, 1986.

SPECIFIC ARTICLES

Furukubo-Tokunaga, K., S. Flister, and W. J. Gehring. Functional specificity of the *Antennapedia* homeodomain. *Proceedings of the National Academy of Sciences* 90 (1993): 6360–64.

Gehring, W. J., Y. Q. Qian, M. Billeter, K. Furukubo-Tokunaga, A. F. Schier, D. Resendez-Perez, M. Affolter, G. Otting, and K. Wüthrich. Homeodomain-DNA recognition. *Cell* 78 (1994): 211–23.

Schier, A. F., and W. J. Gehring. Direct homeodomain-DNA interaction in the autoregulation of the *fushi tarazu* gene. *Nature* 356 (1992): 804–7.

Chapter 9

SPECIFIC ARTICLES

Bellen, H. J., C. J. O'Kane, C. Wilson, U. Grossniklaus, R. Kurth Pearson, and W. J. Gehring. P-element-mediated enhancer detection: A versatile method to study development in *Drosophila. Genes and Development* 3 (1989): 1288–1300.

Bier, E., H. Vaessin, S. Shepherd, K. Lee, K. McCall, S. Barbel, L. Ackerman, R. Carretto, T. Uemura, E. Grell, L. Y. Jan, and Y. N. Jan. Searching for pattern and mutation in the *Drosophila* genome with a P-lacZ vector. *Genes and Development* 3 (1989): 1273–87.

Grossniklaus, U., H. J. Bellen, C. Wilson, and W. J. Gehring. P-element-mediated enhancer detection applied to the study of oogenesis in *Drosophila. Development* 107 (1989): 189–200.

O'Kane, C. J., and W. J. Gehring. Detection *in situ* of genomic regulatory elements in *Drosophila. Proceedings of the National Academy of Sciences* 84 (1987): 9123–27.

Wagner-Bernholz, J. T., C. Wilson, G. Gibson, R. Schuh, and W. J. Gehring. Identification of target genes of the homeotic gene *Antennapedia* by enhancer detection. *Genes and Development* 5 (1991): 2467–80.

Wilson, C., R. Kurth Pearson, H. J. Bellen, C. J. O'Kane, U. Grossniklaus, and W. J. Gehring. P-element-mediated enhancer detection: An efficient method for isolating and characterizing developmentally regulated genes in *Drosophila. Genes and Development* 3 (1989): 1301–13.

Chapter 10

SPECIFIC ARTICLES

Doe, C. Q., Y. Hiromi, W. J. Gehring, and C. S. Goodman. Expression and function of the segmentation gene *fushi tarazu* during *Drosophila* neurogenesis. *Science* 239 (1988): 170–75.

Nose, A., V. B. Mahajan, and C. Goodman. Connectin: A homophilic cell adhesion molecule expressed on a subset of muscles and the motoneurons that innervate them in *Drosophila. Cell* 70 (1992): 553–67.

Salser, S., and C. Kenyon. Patterning *C. elegans:* Homeotic cluster genes, cell fates and cell migrations. *Trends in Genetics* 10 (1994): 159–64.

Chapter 11

GENERAL

Hadorn, E. Transdetermination in cells. *Scientific American,* November 1968, 110–20.

Chapter 12

GENERAL

De Robertis, E., G. Oliver, and C. Wright. Homeobox genes and the vertebrate body plan. *Scientific American,* July 1990, 46–52.

Gehring, W. J., M. Affolter, and T. Bürglin. Homeodomain proteins. *Annual Review of Biochemistry* 63 (1994): 487–526.

Lewis, E. B. Clusters of master control genes regulate the development of higher organisms. *Journal of the American Medical Association* 267 (1992): 1524–31.

McGinnis, W., and R. Krumlauf. Homeobox genes and axial patterning. *Cell* 68 (1992): 283–302.

SPECIFIC ARTICLES

Bachiller, D., A. Macias, D. Duboule, and G. Morata. Conservation of a functional hierarchy between mammalian and insect *Hox/Hom* genes. *EMBO Journal* 13 (1994): 1930–41.

Boncinelli, E., R. Somma, D. Acampora, M. Pannese, M. D'Esposito, A. Faiella, and A. Simeone. Organization of human homeobox genes. *Human Reproduction* 3 (1988): 880–86.

Condie, B., and M. Capecchi. Mice homozygous for a targeted disruption of *Hoxd-3* (*Hox-4.1*) exhibit anterior transformations of the first and second cervical vertebrae, the atlas and the axis. *Development* 119 (1993): 579–95.

Kessel, M., R. Balling, and P. Gruss. Variations of cervical vertebrae after expression of a *Hox-1.1* transgene in mice. *Cell* 61 (1990): 301–8.

Malicki, J., K. Schughart, and W. McGinnis. Mouse *Hox-2.2* specifies thoracic segmental identity in *Drosophila* embryos and larvae. *Cell* 63 (1990): 961–67.

Simeone, A., D. Acampora, M. Gulisano, A. Stornaiuolo, and E. Boncinelli. Nested expression domains of four homeobox genes in developing rostral brain. *Nature* 358 (1992): 687–90.

Van der Hoeven, F., J. Zakany, and D. Duboule. Gene transpositions in the *Hox D* complex reveal a hierarchy of regulatory controls. *Cell* 85 (1996): 1025–35.

Zhao, J., R. Lazzarini, and L. Pick. The mouse *Hox-1.3* gene is functionally equivalent to the *Drosophila Sex combs reduced* gene. *Genes and Development* 7 (1993): 343–54.

Chapter 13

GENERAL

Callaerts, P., G. Halder, and W. J. Gehring. *Pax-6* in development and evolution. *Annual Review of Neuroscience* 20 (1997): 483–532.

SPECIFIC ACTICLES

Halder, G., P. Callaerts, and W. J. Gehring. Induction of ectopic eyes by targeted expression of the *eyeless* gene in *Drosophila*. *Science* 267 (1995): 1788–92.

Quiring, R., U. Walldorf, U. Kloter, and W. J. Gehring. Homology of the *eyeless* gene of *Drosophila* to the *Small eye* gene in mice and *Aniridia* in humans. *Science* 265 (1994): 785–89

Index

Page numbers in *italics* indicate illustrations